World Soils Book Series

Series Editor

Prof. Alfred E. Hartemink
FD Hole Soils Laboratory, Department of Soil Science
University of Wisconsin–Madison
Madison
USA

For further volumes:
http://www.springer.com/series/8915

Aims and Scope

The *World Soils Book Series* brings together soil information and soil knowledge of a particular country in a concise and reader-friendly way. The books include sections on soil research history, geomorphology, major soil types, soil maps, soil properties, soil classification, soil fertility, land use and vegetation, soil management, and soils and humans.

International Union of Soil Sciences

Ferdo Bašić

The Soils of Croatia

Springer

Ferdo Bašić
Faculty of Agriculture
Department of General Agronomy
University of Zagreb
Zagreb
Croatia

ISSN 2211-1255 ISSN 2211-1263 (electronic)
ISBN 978-94-017-7821-3 ISBN 978-94-007-5815-5 (eBook)
DOI 10.1007/978-94-007-5815-5
Springer Dordrecht Heidelberg New York London

Printed on acid-free paper

Springer is part of Springer Science+Business Media (www.springer.com)

Preface

...It is holy truth that for advances in agriculture it is not enough to have a diligent hand only, but knowledge and tuition... because soil as the source of all reaches is a mass of minerals which did not know by itself to adapt to all the needs of mankind. Soil needs a wise guidance... so that all efforts bear fruits and benefit...

From Preface of "Zemljoznanstvo" by Mijo Kišpatić 1877

From its start in the second half of the nineteenth century, soil science in Croatia has a long tradition. Because it began within the framework of agriculture and forestry, for a long time the focus of research was on soil fertility as the dominant soil property due to its function in the production of organic matter, including food, feed, fiber, timber and (more recently) biofuel, in forest and agroecosystems.

In Croatia the forestry profession is dominated by the so-called "Zagreb forest school", among the first of its kind in Europe, studying "non-productive"—so-called "public good"—functions of forest and the influence of the silvi-ecosystem on other spheres—atmosphere, hydrosphere and biosphere, or terrestrial and semiterrestrial ecosystems. The development of an integrated holistic approach to sustainable development and spectacular advances in soil science and soil chemistry, particularly in regard to analytical methods and laboratory equipment for soil survey, made possible critical evaluation of the very complex impact of soil as well as other components of the agroecosystem on other ecosystems and the environment, especially the biosphere. This integrated holistic approach undoubtedly claims that soil management should be equal in importance to life management. With this knowledge in mind, the radical reform of European higher education known as the "Bologna process" has resulted in a changing of traditional names of large and prominent European universities of agricultural and forest science in Vienna, Prague and other European centers into "Universities (Studies) of Life Sciences". What a pity, as these terms, when translated into the Croatian language, give an inappropriate name.

Croatian soils have been explored, presented in soil maps, and described in regional monographs, but a single monograph on Croatian soil has not been published until now. Therefore, after becoming an independent state, there was a logical need to present Croatia's soil resources, with adequate attention being paid to multifunctionality and "non-productive functions" of soil, which are of particular economic, social, and environmental importance. This monograph should perform this role and encourage the production of an optimal sustainable land (soil) management system for all agroecological conditions in Croatia.

Croatia is unique in that of all the states of Europe it is the first whose soil monograph will be published in a foreign language—in English. As this book will be published shortly before Croatia joins the European Union, its important role is to inform the professional public of this national Croatian resource. As with any other book only time will tell to what extent that intention will succeed. In this regard, any remarks on the text from readers of this monograph are welcome.

Acknowledgments

Most of the data and knowledge on the pedosphere of Croatia presented in this book have been collected from the beginning of the 1970s in the second half of the last century. For my generation of Croatian soil scientists, the inspiration and motto for their research activities was the inspirational message (cited in the book) of M. Gračanin (1901–1981), the leading soil scientist of Croatia, who aimed soil science in our country to "genetically school". His very pertinent advice and recommendations of basic, lastingly valid principles of academic work and life remain in my mind.

Upon completion of the book it is my pleasant duty to thank to all participants, from the idea and start to publication of the book. First of all, I address my thanks to Dr. Alfred E. Hartemink and Dr. Robert Doe who initiated the publication of this series of books, and encouraged me in all phases of the project. My thanks go to all the personnel of Springer Verlag, particularly Dr. Elodie Tronche—Assistant Editor of Springer Verlag from Dordrecht, The Netherlands—for excellent, sensitive, and highly professional help and patience during the preparation of this manuscript. My English is not as fluent as I would like, and without their help this text would be modest. For the same reason thanks go to the personnel of Scientific Publishing Services in Chennai, India.

Special gratitude is addressed to my friend A. Vranković, lecturer on soil science in the Faculty of Forestry University of Zagreb, for support at the decisive moment—the crossroads of my academic career. Unforgettable is his evocative discussion, informing me on genesis, properties, and desirable management of soils on karst—limestone and dolomites—and the distribution of the rhizosphere of forest vegetation in these soils.

The same thanks I address to N. Vuletić, former director of the Institute for Tobacco and an excellent expert in soil chemistry.

Very valuable are experiences in research into forest ecosystems with S. Matić, member of Croatian Academy of Sciences and Arts and leading Croatian scientist in sustainable, environment- and nature-friendly forest management. From the same circle, the best memory remains also with the late Prof. Emeritus of the Faculty of Forestry University of Zagreb, B. Prpić with special concern for soil–plant relationships and the state of the rhizosphere as an indicator of soil as site component offorest ecosystems of hydromorphic soils.

With cordial thanks I address all members of the team of soil scientists engaged in a scientific project started in the former Yugoslavia: the General Soil Map of Croatia at a scale of 1:50,000, as the main basis of this work; firstly M. Bogunović as the leading cartograph of Croatian soil cartography, from the beginning (1969) to the end of this project (1986), followed by J. Martinović, Ž. Vidaček, I. Šimunić, M. Adam, B. Miloš, V. Pavlić, brother of my I. Bašić, the late A. Čolak, B. Mayer, P. Rastovski and I. Šalinović. Each of them during the period 1971–1986 participated in rich discussions of open soil profiles all over Croatia, enriching my knowledge of soils of Croatia and experiences in soil cartographic work.

My cordial thanks go to Prof. Husnjak, Head of the Department of Soil Science, Faculty of Agriculture, University of Zagreb and member of European Soil Bureau Network (ESBN), for his constant support regarding maps, figures, and photos, but also for the manuscript of *Soil Taxonomy*, now in print but published for the first time in this book. In the same way I am grateful to my colleagues I. Kisić, M. Mesić, B. Komesarović, B. Vrbek, N. Pernar and Ž. Vidaček for providing some photos and/or data on soils.

For encouraging my first steps within "the world of knowledge" on soil (land) management, I owe gratitude to the late V. Mihalić the founder of the "Zagreb school of land management" as a school of holistic admission to soil fertility, one of the founders of the International Soil Tillage Research Organization (ISTRO). I address the same gratitude to the leading scientist of this school, A. Butorac, especially in regard to field experiments as an important basic method for examining hypotheses in land management. My activities followed his thesis that soil type and its properties on a concrete plot of land have to be in the focus of decision-making concerning an efficient sustainable land management practices, especially tillage, fertilization, liming and maintenance of quality and quantity of humus in the soil. Therefore, there is no reason for a uniform "always-valid" solution.

Because each plot of land consists of more systematic soil units, the adaptation of land management systems to soil type as the most complex component of agroecosystem and agroecological conditions is the only right way… as is stressed in the title of the book (Adaptable tillage) of my friend of M. Birkas (University of Godollo)—mistress of management of Hungarian soils!

Also, I would like to express my deep gratitude to the "dean of all deans" of the Faculty of Agriculture, University of Zagreb, F. Tomić, member of Croatian Academy of Sciences and Arts, the leading scientist in the field of soil hydroamelioration, for experiences exchanged during our co-operation in studies related to agroecological conditions and guidelines of development for agriculture in several counties of Croatia.

Of great importance were discussions and results of long-term joint research with my friend O. Nestroy (Graz, Austria), especially on soil genesis, as well as G. Varallyay (Member of Hungarian Academy of Sciences and Arts), M. Džatko (Bratislava, Slovak Republic), H. Resulović, M. Vlahinić and H. Čustović (Sarajevo, Bosnia and Herzegovina). Of great importance was the support of Prof. W. Blum (BOKU, Vienna)—former secretary of ISSS, IUSS and founder of the association of European soil scientists (ESSS).

Also very valuable also were my experiences of working in the European Soil Bureau Network (ESBN) and its Steering Committee under the chairmanships of L. Montanarella, A. Jones, A. Arnoldusen, and R. Creamer, as well as the very rich collection of published scientific papers and books of ESBN, especially *Soil Resources of Europe and Soil Atlas of Europe*.

Finally, with exclusive importance I would like to address special warm thanks to my long-term close associates I. Kisić and M. Mesić, younger collaborators Ž. Zgorelec, I. Šestak, A. Jurišić and other staff members of the Department of General Agronomy of the Faculty of Agriculture of the University of Zagreb. All of them very successfully continue to keep the tradition of the "Zagreb school of land management and soil protection" and participated in an un forgettable, creative, serious, academic working atmosphere during long-term joint work and fruitful research into very complex problems of soil management, soil degradation and protection, as a basic precondition to sustainable land use.

The last but really the warmest words of acknowledgement I address to my family, as the main driving force and inexhaustible source of positive energy, understanding, and tenderness, firstly to my wife Ankica who (as agronomist) supported and encouraged my whole life-work, my son Tomislav (grad. eng. of oil mining), his wife Sandra and children (my grandchildren) Domagoj and Luka, as well as my daughter Marijana (grad. oecc.). Each of them in their own way initiated a creative and stimulative atmosphere and cheered up my work.

Zagreb, September 2012

Author
Dr.sc Ferdo Bašić, Professor Emeritus
Faculty of Agriculture, University of Zagreb
Zagreb, Croatia

Contents

Introduction

1

Croatian people, in a real sense, 'live on the soil'; on their own part of the pedosphere they built their own life in the past, and will build it in the future. The soils of Croatia are the greatest wealth of the Croatian people, an inexhaustible source of their vigour and the foundation of the Croatian homeland....

Prof.dr.sc. M. Gračanin (1942).

1.1 General Information Regarding Croatia

Located on the crossroads between North and South, East and West, Croatia was, and remains, a bridge and bond between the various cultures of western civilization, and has always been a maritime nation, communicating with the whole world.

By its geographical position Croatia is a Central European and Mediterranean country, located at the crossroads between central Europe and the Mediterranean. In the west, Croatia borders on Slovenia, to the north Hungary, to the east and south Serbia and Bosnia and Herzegovina, and to the south Montenegro. It has a sea border with Italy to the west. Figure 1.1 shows its geographical position.

With its area of 56,542 km^2, Croatia is a European country of medium size, similar to Denmark, Estonia, Latvia, Lithuania, Slovakia, Bosnia and Herzegovina, the Netherlands, and Switzerland. According to the latest census from 2011 the population of Croatia numbered 4,290,612, living in a land territorially organized into 21 counties, 124 towns and 426 municipalities. According to population, Croatia is similar to Denmark, Lithuania, Bosnia and Herzegovina, Finland, Ireland, Moldova, Norway, and Slovakia. On the basis of population 24 European countries (out of 43) are larger than Croatia. The gross domestic product in 2011 amounted to 10,394 EUR per capita. Basic geographic data are presented in Table 1.1.

The Croatian area is divided into three natural parts according to geomorphological, lithological, and agro-ecological characteristics.

- **The Pannonian and Sub-Pannonian area** comprises lowland and hilly parts of eastern and western/north-western Croatia, which (except for Baranja) are bounded by three rivers that are praised in the hymns of the state: the Sava, Drava, and Danube Rivers. The rivers have numerous tributaries from Croatia and neighboring countries, and belong to the Danube–Black Sea catchment area, Europe's largest. The area has traditionally developed agriculture—the growing of food crops and livestock, and freshwater fisheries. Slavonia and Baranja in the east are most suitable for growing grain; moist valleys and mountain regions are rich in forests, and the north-western part, which centers on Zagreb, has developed industry.
- **Mountains** separate Pannonian Croatia from the Adriatic (Mediterranean). They are sparsely populated and economically underdeveloped, but have national parks preserving nature in protected areas including clean rivers with an abundance of drinking water. Their future development will be based on forestry and timber. With appropriate rural development there will be particularly favorable conditions for organic farming, livestock breeding, freshwater fishing, winter and agro-tourism.
- **Adriatic Croatia** comprises a narrow coastal belt separated from the hinterland. It is a karst area with hot and long summer periods without rainfall. A few streams are truly beautiful rivers, with picturesque canyons and preserved untouched beauty. The Croatian Adriatic coast is divided into a northern (Istria and Kvarner) and southern part (Dalmatia) with a well-defined longitudinal division of island, coastal, and hinterland areas, each of which has its own peculiarities. In the karst area, which is dominated by a "green oasis", are fertile karst fields.

A characteristic of Croatia is the excellent transportation links over modern highways, so that practically all destinations are quickly reached by visitors. The economic development of Croatia is founded upon three branches of the economy—agriculture, forestry, and tourism, which are all environmentally sustainable, so that Croatia sees itself as in a state designed for sustainable development.

Croatia's Adriatic coast is one of the most developed in Europe. By area the largest island is Krk (410 km^2), followed by Cres (406 km^2), Brač (395 km^2), Hvar (300 km^2), Pag (285 km^2), and Korčula (276 km^2). Croatia has signed several international conventions, which obligate it to preserve its natural riches. According to the Act on Nature Protection, 433 areas, with a total area of 515,093 ha, are being protected to a lesser or greater degree. Most beautiful are the national parks and nature parks shown in Fig. 1.2.

F. Bašić, *The Soils of Croatia*, World Soils Book Series,
DOI: 10.1007/978-94-007-5815-5_1, © Springer Science+Business Media Dordrecht 2013

Fig. 1.1 Geographical position of Croatia

Croatia has eight national parks, with a total area of 99,804 ha: Plitvička jezera (29,489 ha), Paklenica (10,200 ha), Risnjak (6,400 ha), Mljet (5,480 ha), Kornati (23,400 ha), Brijuni (3,635 ha), Krka (10,300 ha), and Sjeverni Velebit (10,900 ha). Croatia is situated between 45°29″and 46°33″ north latitude (NGL) and 13°30″ to 16°22″ east longitude (EGL). The geographical coordinates of the extreme points of the territory are shown in Table 1.2.

The independent Croatian state was formed in 1990 after the disintegration of the former Yugoslavia and it received international recognition in 1992. 2013 is the expected date of EU accession, after which Croatia will seek further development within the community of European nations.

1.2 Short History of Soil Science and Soil Survey in Croatia

Soils were studied by the scientists of Croatia long before the recent advent of soil science. The first data on soil were collected within the fields of geology and mineralogy. After

Table 1.1 Some geographical data of the Republic of Croatia

Total area	**87,661 km^2**
Land area	56,594 km^2
Territorial sea and internal waters	31,067 km^2
Coastline	**5,835.3 km**
Coastline—land only	1,777.3 km (30.5 %)
Coastline—islands	4,058 km (69.5 %)
Total economic area of the sea	**113,680 km^2**
Length of land boundary 2,028 km	**2,028 km**
Length of land borders with Slovenia	501 km
Length of land borders with Bosnia and Herzegovina	932 km
Length of land borders with Hungary	329 km
Length of land borders with Serbia	252 km
Length of the river border with neighboring countries	**About 1,600 km**
Length of sea border with Italy	**About 900 km**
Number of islands, islets, and reefs	**1,246 (47 inhabited); 389 and 78**

Fig. 1.2 National parks (*green*) and nature parks (*yellow*) of Croatia

Table 1.2 Geographic coordinates of extreme points of Croatian territory

	Settlement	City	County	NGL	EGL
N	Zabnik	St. Martin on Mura	Međimurje	46°33″	16″22′
S	Island Galijula (Palagruža)	Komiža	Dubrovnik-Neretva	42″23′	16″21′
E	Ilok-Radjevac	Ilok	Vukovar-Syrmium	45″12′	19″27′
W	Bašanija-cape Lako	Umag	Istrian	45″29′	13″30′

Dokutchaev's monograph on the Russian chernozem, the next decisive event in the development of soil science was the first conference on European Agrogeology, attended by prominent geologists from 23 countries, held in Budapest in 1909 (Photo 1.1). The conference was attended by several followers of Dokutchaev: Glinka, Ramman, and Atterberg. The famous Croatian geologist Gorjanović Kramberger, far better known for finding fossils of Neanderthal man in Krapina, also attended. During the 100th Anniversary events in September 2009, members of the European Soil Bureau Network (ESBN) held a special session and toured localities that had been used for discussions in the 1909 conference, representing the infancy of soil science.

The third conference held in Prague 1922 was renamed in the Conference of Pedology, while the fourth conference, the second under the banner of Pedology held in Rome 1924 established the International Society of Soil Science—ISSS (Micheli 2009). Today we have the International Union of Soil Science (IUSS) as the global organization with thousands of members on all continents. The main activity is the congress of IUSS held every four years. The last (the 19th Congress) was held in Brisbane in 2010, and the next will be in 2014 in Seoul, South Korea (www.iuss.org).

European soil scientists are organized into national societies as autonomous organizations. Following a proposal by the famous Austrian soil scientist and secretary general of ISSS, Prof. W. Blum, in 2006, these societies were associated into the European Confederation of Soil Science Societies (ECSSS). The objectives of the ECSSS are to foster collaboration and cooperation amongst the national societies of soil science in Europe and amongst

Photo 1.1 Participants of the first conference of Agrogeology—Budapest 1909

European soil scientists in all branches of the soil sciences and their applications, and to give support to the above in the pursuit of their activities. As part of its aim of facilitating the organization of meetings and conferences, its principal meeting is the EUROSOIL congress organized in a four-yearly cycle. The first was in Vienna in 2008 and the second followed in Bari in 2012.

Croatia has a special interest in the activities of the Joint Research Centre (JRC) of the European Commission in Ispra, Italy. As a service of the European Commission, the JRC provides scientific and technical support for the conception, development, and implementation and monitoring of EU policies, including its soil policy. The European Soil Bureau Network (ESBN) was created in 1996 as a network of national soil science institutions, managed through a permanent Secretariat located at the JRC. The creator, initiator, and head of activities of ESBN is Luca Montanarella (the author of this publication was a member of ESBN from the beginning until his retirement in 2010). Since October 2000, the ESBN has been part of the Land Management Unit of the Environment Institute (IES), one of four institutes at the JRC Ispra site. The ESBN's aim is to carry out scientific and technical work programs in order to collect, harmonize, organize, and distribute soil information relevant to Community policies to a number of Directorates General (DGs), to the European Environment Agency (EEA), and to individual Institutions of the EU Member States. The origins of the ESBN go back more than a decade and are inextricably linked with the compilation of a European Soil Map and associated attribute databases (Toth et al. 2007). An EC Soil Map was produced at a scale of 1:1,000,000 in the 1970 s by a loose network of academic soil scientists.

The European Soil Portal contributes to a thematic data infrastructure for soils in Europe and presents data and information regarding soils at a European level. It connects to activities within the JRC concerning soil. It also serves for promotion of activities of the ESBN. The European Soil Portal currently contains a large amount of soil data, maps, information, atlases and applications; most of the data offered are at a European scale, while, when possible, links to national or global datasets are provided (http://eusoils.jrc.ec.europa.eu/).

The most important of these activities was the Soil Atlas of Europe (2005), produced by A. Jones and L. Montanarella.

The role of the ESBN is to recognize the main environmental problems and questions originating from soil, which include: leaching of different organic and inorganic compounds and agrochemicals (nutrients, nitrates, residues of pesticide), heavy metals, waste disposal, degradation of the structure of the soil through loss of organic matter, risk of erosion (by water and wind), immobilization of radionuclides, water supply, assessing the suitability and sustainability of traditional and alternative crops, etc (Van-Camp et al. 2004). Soil activities within the JRC are concentrated in a specific JRC Action, named FP7 "Soil Data and Information Systems." This action, which belongs to the Land Management and Natural Hazards Unit of the Institute for Environment and Sustainability, will run throughout the period 2007–2014 and has a number of objectives, including the establishment of the European Soil Data Centre (ESDAC) as a single focal point for all soil data and information in Europe, and also research and development of advanced modeling techniques, indicators and scenario analyses in relation to the major threats to soil: erosion by water and wind, decline of organic matter content, compaction, salinization, landslides, sealing, contamination, and loss of soil biodiversity. Of special importance for Croatia is the technical assistance in negotiating the thematic strategy for soil protection and the proposal for a soil framework directive through the EU institutions and their subsequent implementation at community and member state level.

Starting and for a long period developing within agricultural and forest sciences, in the last decades research activities in soil science have become more and more occupied by the "non-productive" functions of soil: as acceptor, accumulator, transformer, or source of emission into the atmosphere, hydrosphere, and/or biosphere (food chain) of different environmentally relevant elements and/or compounds. A logical consequence is for soil science to "move" to the category of environmental sciences.

A milestone in the development of soil science in Croatia occurred during the Austro-Hungarian Monarchy. Namely, after the establishment of the Higher Royal Agricultural and Forestry School at Križevci in 1877, M. Kišpatić published *Zemljoznanstvo* (Earth-knowledge), the first textbook on soil science in the Croatian language, and one of the first such books in the world. This was followed by the establishment of the Department of Soil Science, Petrology and Mineralogy at the Forest Academy in Zagreb in 1910, which was later integrated into the Faculty of Agriculture and Forestry of the Kingdom of Yugoslavia, established in 1919 in Zagreb. The Croatian soil scientists Šandor and Moskovič accepted the "Russian genetic school," a decision strongly influenced by the excellent textbook *Agropedology* by the prolific pedologist Alexander Stebut—a professor of soil science of Russian origin at the University of Belgrade (Stebut 1949, 1950, 1953).

Indisputably, the famous soil scientist, plant physiologist and plant ecologist Mihovil Gračanin (1901–1981)—professor at the Faculty of Agriculture in the University of Zagreb—had a most important role to play in the development of soil sciences in Croatia (Gračanin 1951). His message, printed under the title of this chapter, inspired a generation of Croatian soil scientists and became a motto for their activities. His epochal textbooks, printed in Zagreb, had a crucial role in the education of a generation of students and professionals in Zagreb and Skopje. He was followed by A. Škorić, a long-term professor of Pedology in the Agricultural Faculty, President of the Project Council of the Soil Map of Croatia 1:50,000, and leading author of regional monographs on Slavonia and Baranja (Škorić et al. 1977), Istria (Škorić et al. 1987) and Mountainous Croatia (Škorić et al. 1992). As a leading cartographer of the Soil Map of Croatia and author of the first soil maps of an independent Croatia, M. Bogunović, professor of Pedology of the Faculty of Agriculture, who participated in preparing the General Soil Map of Croatia from the beginning to the final stage, left a lasting impression.

The first form of organization of Croatian soil scientists was the Yugoslavian section of the International Society of Soil Science. The first president of this section in the period 1931–1940 was M. Gračanin. We therefore take 1931 as the year of establishment of the Croatian Society of Soil Science—CSSS. After World War II the CSSS continued its activities within the Yugoslavian Society of Soil Science, as a member of ISSS (Bašić et al. 1994). In 1891 the first laboratory for soil analysis was founded in Zagreb. After Croatia became an autonomous state, the CSSS continued its activities as an autonomous scientific society. The most recent congress of the society, the 11th, was held in 2011.

The most significant, epoch-making document on Croatian soils is the General Soil Map (GSM) of Croatia at the scale of 1:50,000 prepared in the period 1964–1985. The project started in 1965, as a federal project of Yugoslavia, with a standard, uniform and subsequently used method of cartography, revision, and printing of maps. In the former state the project was named after the name of the Republic, so for Croatia it was the General Soil Map of Croatia (GSMC); it was at the scale 1:50,000. This priceless document collected analytical data on about 6,000 soil profiles and is the most important actual document of Croatian soils. It was made using the most up-to-date methods of that time, with application of aerial stereo-photographs, while the soils are presented on printed sheets of the soil map at 1:50,000 (Kovačević and Jakšić 1964). The print quality is high, considering the time. This map, presented on 186 sheets, contains data on soil properties and the spatial distribution of soils in Croatia, collected with an observation density approximately amounting to one soil profile per 1,000 ha of land. Figure 1.3 shows the printed sheets and sheets in manuscript form, awaiting printing.

The preparation of individual sheets of the GSM was followed by the creation of regional soil monographs. Four monographs with maps have been published to date: *Soils of the Upper Sava River Valley* (Kovačević et al. 1972), *Soils*

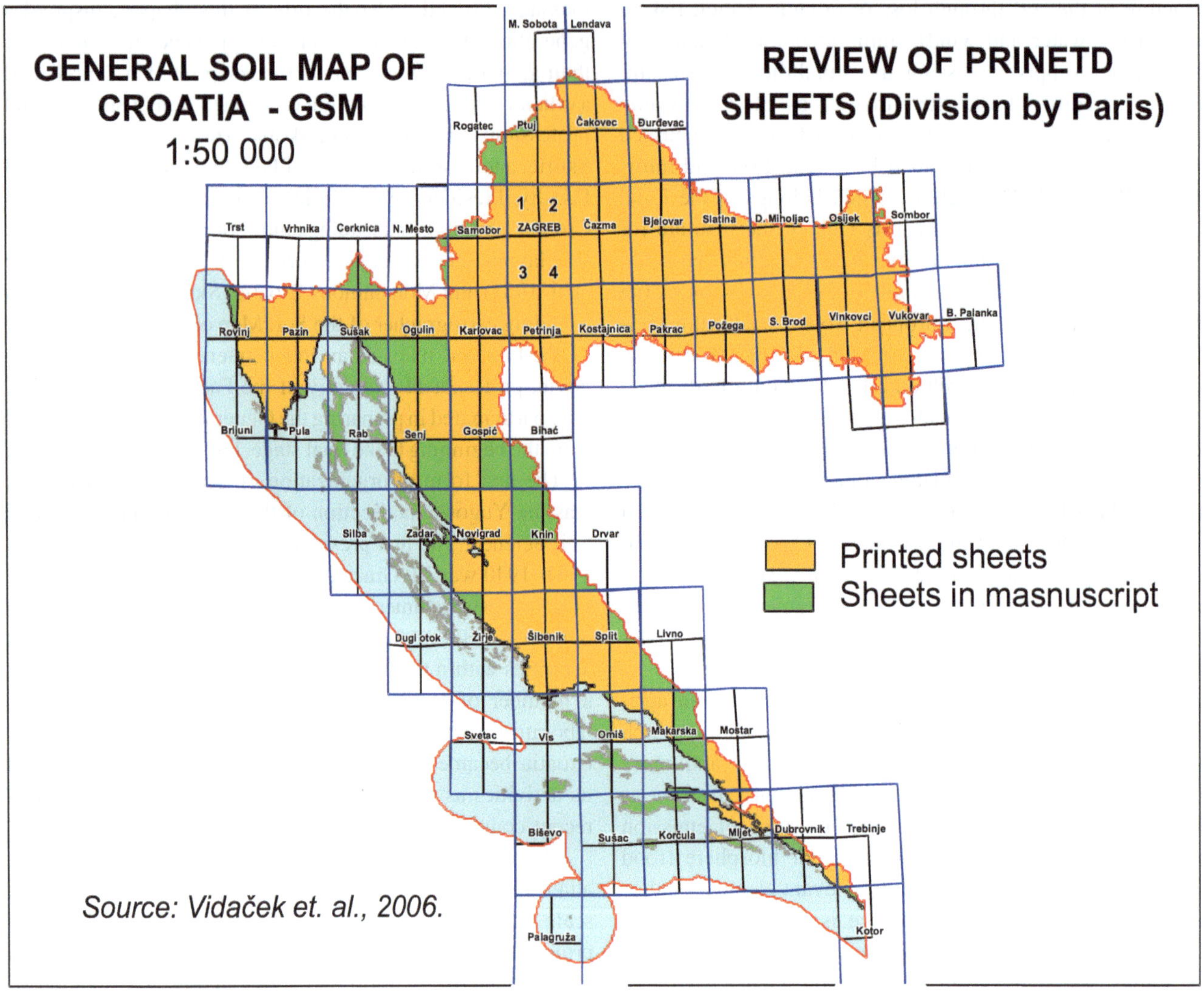

Fig. 1.3 General soil map of Croatia 1:50,000 and an outline of the 186 sheets of the GSMC

of Slavonia and Baranja (Škorić et al. 1977), *Pedosphere of Istria* (Škorić et al. 1987), and *Soils of Mountainous Croatia* (Škorić et al. 2003).

The GSMC data form a reliable basis of a unique information system on the soils of Croatia—the Croatian Soil Information System (CROSIS)—which is a part of the Environment Information System of Croatia (EISC). CROSIS forms an important basis for the creation of sustainable soil and land management practices in agriculture and forestry as well as the Croatian Soil Protection Strategy (CSPS), and has other uses related to multifunctionality of soil/land and development of a thematic strategy for soil protection in the EU.

All activities, from initiation to preparing and printing a Manual of Methodology for the production of the Soil Map of Yugoslavia 1:50,000 (Kovačević and Jakšić 1964), followed by the first testing of the usability of this Manual in the field, were carried out by the Institute for Soil Science and Soil Technology in Zagreb. The very rich GSMC documentation, which, along with printed material, also contains unique manuscripts, is kept at the Department of Soil Science, Faculty of Agriculture, which was the ultimate center of cartographic activities in Croatia. The data await up-to-date computer processing (digitalization), and there is no doubt that they represent a more than solid and reliable basis for a unique information system on the soils of Croatia.

Today it is necessary to begin preparing a Revised General Soil Map of Croatia—RGSMC, using the possibilities provided by the modern technical solution of multispectral satellite images. Revision of the GSMC is also justified by the fact that new soil survey material and arguments have been collected from the time the first sheets were prepared in 1964 up to its completion in 1985. In addition, the new criteria for international classification, correlation and communication—the World Reference Base for Soil Resources (WRB 2006) seeks amendment of the classification of Croatian soils and its use in the RSMC, simply for ease of communication, something that is often lacking in the history of Croatian soil science (Bašić 2007).

Several soil maps of smaller scale were prepared for the area of the former Yugoslavia, mostly for educational purposes. The work of Croatian authors Škorić and Bogunović (1977) is worth pointing out: The Soil Map of Yugoslavia, 1:2,000,000, was based on a map prepared for the needs of the Soil Map of Europe, initiated by the FAO at a scale of 1:1,000,000. Using the same maps and soil survey material, Bogunović (1992) prepared the Soil Map of Croatia at the scale of 1:1,000,000, the first one for the independent state.

Conversely, environmental protection is unimaginable without a soil map, and since its planning is digitally based, the GSMC documentation (Bogunović et al. 1996) should be used to prepare a Digital Soil Map separately for each county of Croatia.

References

Bašić F, Bertović S, Martinović J, Miljković I (1994) Spomenica, uz 90. obljetnicu rođenja i 10. obljetnicu smrti Mihovila Gračanina, Izdvojeni tisak Agronom. glasnika br. 1-2,/94, str. 228, Zagreb

Bašić F (2007) Soils of Croatia—status—problems and anticipated solutions. In: Proceedings of the Mediterranean Conference on the status of Mediterranean soil resources: actions needed to promote their sustainable use. Tunis, p 87–111

Bogunović M (1992) Pedološka karta Hrvatske 1: 1 000 000. Agronomski fakultet Sveučilišta u Zagrebu, Zavod za pedologiju, Zagreb

Bogunović M., Vidaček Ž, Racz Z, Husnjak S, Sraka M (1996) Namjenska pedološka karta Republike Hrvatske, mjerilo 1:300 000 (komentar i uputa za digitalno korištenje), Agronomski fakultet Sveučilišta u Zagrebu, Zavod za pedologiju, str. 46, Zagreb

Gračanin M (1951) Pedologija—III dio—Sistematika tala, udžbenik Sveučilišta u Zagrebu, Školska knjiga, str. 298, Zagreb

Kovačević P, Jakšić V (1964) Priručnik za terenska pedološka istraživanja, str. 72, Sarajevo

Kovačević P, Kalinić M, Pavlić V, Bogunović M (1972) Tla Gornje Posavine. Institut za pedologiju i tehnologiju tla Poljoprivrednog fakulteta Sveučilišta u Zagrebu, Zagreb

Micheli E (2009) Retrospect of the 1st international conference of agrogeology, presentation on conference bridging the centuries conference, Hungarian Soil Science Society Department of Soil Science and Agrochemistry, Hungarian Academy of Science and Arts, Budapest

Stebut A (1949, 1950, 1953) Agropedologija I, II i III dio, Beograd

Škorić A, Bogunović M (1977) Pedološka karta Jugoslavije 1:2000 000. Zavod za pedologiju Agronomskog fakulteta, Zagreb

Škorić, A i sur (1977) Tla Slavonije i Baranje. Projektni savjet za izradu Pedološke karte Republike Hrvatske, str. 256, Zagreb

Škorić, A i sur (1987) Pedosfera Istre. Projektni savjet za izradu pedološke karte Republike Hrvatske,str. 92, Zagreb

Škorić A, i sur (1992) Tla gorske Hrvatske, Zavod za pedologiju Agronomskog fakulteta, rukopis, str. 340, Zagreb

Škorić A i sur (2003) Tla gorske Hrvatske, monografija, Ministarstvo zaštite okoliša i prostornog uređenjag fakulteta, str. 169, Zagreb

Toth G, Stolbovoy V, Montanarella L (2007) Soil quality and sustainability evaluation, an integrated approach to support soil-related policies of the EU—a JRC position paper. Institute for Environment and Sustainability, Office for Official Publications of the EU, Luxemburg, p 40

Van-Camp L, Bujarrabal B, Gentile AR, Jones RJA, Montanarella L, Olazabal C, Selvaradjou S (2004) Reports of the technical working groups established under the thematic strategy for soil protection: Vol I—Introduction and Executing Summary, p 126, Vol II—Erosion, pp 127–309, Vol III—Organic matter, pp 31–496, Vol IV Contamination and Land Management, p 497–621, Vol V Monitoring, pp 653–718, Vol VI Research, Sealing and cross-cutting issues, p 719–872. Office for Official Publications of the EU, Luxemborgh, p 94

WRB (2006) World reference base for soil resources World soil resources reports 103. Food and Agriculture Organization of the United Nations, Rome, p 128

Websites

European Soil Portal http://eusoils.jrc.ec.europa.eu/

Worldometers http://www.worldometers.info/

2 Soil as a Natural Resource of Croatia

… if Moses had foreseen what suicidal agriculture would do to the land of the holy earth, might he not have been inspired to deliver another Commandment to establish man's relation to the earth (soil) and to complete man's trinity of responsibilities to his Creator, to his fellow men and the Holy Earth… what has been called the "Eleventh Commandment".

W.C. Lowdermilk (1994).

2.1 Soil as a Unique Natural Body

An incredibly complex array of elements and compounds flows in balanced fashion, into our bodies along the nutrient chain. We are what we eat. In a sense we are unique, moist packages of animated soil.

F.D. Hole (1988)

Soil (pedosphere) is a natural, terrestrial body (sphere), a thin surface layer placed between the parent rock (lithosphere) and the atmosphere. The mineral component of soil is generated by processes of soil genesis: destruction and synthesis, removal and accumulation through the combined influences of water, air, temperature and different micro- and macro-lifeforms (flora and fauna). As all the factors of soil genesis in Croatia are highly varied in space and time, the soils of Croatia are very heterogeneous. The distribution of various soils depends on geomorphology—relief and topography, physical properties and chemical composition of parent rock, climatic conditions—daily and annual temperature variations, quantity, distribution and form of precipitation (leaching or accumulation of salt). A great influence on the distribution of soil types is plant cover, and to a greater or lesser extent sometimes the decisive influences are anthropogenic ones (Bašić 1976, 1982, 2009; Resulović et al. 2008).

In soil as a natural, terrestrial milieu, plants as autotrophic organisms originate and it is where the life of all organisms, including heterotrophic organisms, ends. As a natural body, soil is a more complex medium than the other members of the "ecological triad"—air and water. It may indeed be the most complex and dynamic system known to science. According to one theory the word "chemistry" is derived from "al-kīmīā," the ancient Egyptian name for Egypt (*khem*, *khame*, or *khmi*), meaning "black soil" of the Nile River, in contrast to the surrounding sandy desert (Zgorelec et al. 2012). Composed of mineral and organic, solid, liquid, and gaseous components, it contains large numbers of living organisms and plant roots and it is the medium which supports life in its broadest sense. Soil produces and contains all of the elements necessary to life, including air and water, which are movable systems, but the pedosphere is site-fixed and specific, and more stable than the other spheres. However, soil is a living system and it shows great variability in space and time (evolution) (Bašić and Franić 2003).

Soil genesis is a very slow process. The slowness of soil genesis reflects the fact that, as it says in Montanarella (2007), the formation of 2-cm layer of soil in nature needs 500 years. The formation of a 30-cm-thick soil layer requires 1,000–10,000 years—for example chernozems, Croatia's most fertile soil around Ilok in western Syrmium so soil should be considered nonrenewable. A substrate widespread in Croatia is limestone and this is the slowest in terms of soil formation. Durn (2003) states that the formation of a 50-cm layer of red soil (terra rossa) around Umag in Istria takes up to 2 million years, while in that time 50-m-high cretaceous limestone cliffs are dissolved. One can lose this layer if it is exposed to torrential rain for only one day!

Compared to water, air or biota, soils are not known to humans, because they are hidden under their feet and they cannot use them in the way they use air, water, or biota. Soil is a very complex, living system, which has all the characteristics of an organism: it emerges, develops from stage to stage, and is therefore subject to evolution (soil evolution), whose speed depends on the lithological conditions—the parent material from which the mineral soil is derived and is generally faster on loose, but slower on hard rocks. As current users of the land, we actually find different stages of soil evolution on different parent materials. Although the soil is a renewable natural resource because it is constantly produced, due to its slowness of formation it is conditionally renewable, so Varallyay (2000, 2005) tentatively classified it as a renewable resource.

Confirmation of the value and importance of soil is deeply engrained in the minds of Croatian men. Photo 2.1 shows the karst landscape near Primošten city in Dalmatia, which can be seen today at the UN headquarters in New York.

F. Bašić, *The Soils of Croatia*, World Soils Book Series,
DOI: 10.1007/978-94-007-5815-5_2, © Springer Science+Business Media Dordrecht 2013

Photo 2.1 Vineyard with "dry walls" typical of the Croatian karst—the "dry walls" built from the stone dug up during the clearing of the soil—historical testimony of the persistence and energy of domestic man in the fight for survival. There is truth in the poetic words that the soil is wetted by the sweat of the Dalmatian vine grower

As the pedosphere is a "sphere of interaction of all spheres" and as they are interdependent we support the ideas of Varallyay (2005) on equal treatment of soil with the other members of the "ecological triad": water–air–soil. It is evident that all these goods are equally important, but the soil (land) is the basis of one's personal property.

Many of the problems are generated by inappropriate land management at the "local level," and the consequences are reflected regionally, continentally and, ultimately, at the global level. According to historical and social policy acquis, modern civilization property rights are unquestionable, land is private property, but the problems generated by wrong land management are public ones (Urushadze 2002).

In Croatia there are 18 million land parcels situated on agricultural land representing the property of over 450,000 farm households, companies, and individual owners who do not live by agriculture, but their property right is unquestionable (Bašić 2006, 2007). It remains to be seen how and whether it is possible to monitor the management of these soils, and turn this management in the desired direction, thereby reconciling the rights derived from private property on the one hand and public interest on the other.

2.2 Soil Functions: Roles of the Croatian Pedosphere

As soil science as an autonomous natural science began and for a long time developed within the agricultural and forest sciences, the focus of interest was yield, which means soil fertility and its physical, chemical, and biological components. Yield was one of the criteria for land valuing and soil classification. The aim was usually maximal yield, with maximal use of machinery and consumption of agrochemicals including biocides. Under the pressure of data on the hunger of populations, in the center of interest of the world food market was food security, meaning the food production function of soil. All other functions were marginalized or ignored. Trends toward more intensive and specialized farming systems have very successfully increased the ability of mankind to ensure food security and feed the world, but, in some cases, at the expense of social and/or environmental goals (Blum 2002). One of the consequences of such a one-sided orientation was degradation of the environment—water (hydrosphere), soil (pedosphere), air (atmosphere)—as well as problems with food safety (BSE, *Escherichia coli* outbreaks). The agricultural scientists and decision makers should achieve an optimal balance between social, environmental, and economic objectives on the national, regional, continental, and global levels (Birkas et al. 2008).

Step-by-step with the affirmation of sustainability, well-known non-food functions of soil move into focus. Agricultural activities have evolved from a period of economically acceptable yield, to sustainable yield and today socially, economically, and environmentally sustainable yield of biomass for food, feed, timber, and biofuels. While sustainable agriculture is based on long-term goals and not a specific set of farming techniques, it is usually accompanied by a reduction of input in favor of managing on-farm resources. A good example is the reliance on symbiotic and/or nonsymbiotic nitrogen fixation from legumes instead of use of commercial nitrogen fertilizers. Low-input agriculture is only one of several alternative farming systems that are adaptable to sustainable agriculture. Low-input farming is based on a reduction, but not elimination, of commercial agrochemicals like fertilizers and pesticides. Croatian farmers accepted this opportunity primarily motivated by reduced costs, only very rarely to minimize any negative impact on the environment. Let us analyze the food and non-food functions of soils in Croatia, as appropriate policies in support of food security and safety, land tenure security, soil and water protection, and rural development are key elements for sustainability at the national level (Bašić 1998; Bašić and Franić 2003; Bašić et al. 2003).

2.2.1 Primary Production of Organic Matter: Soil is a Source of Food, Feed, Timber, and Biofuels!

The most important, irreplaceable, and primary function of soil is its productive function, its position as the initial and final link in the chain of bio-transformation, and its role in supplying plants with water, air and nutrients for photosynthesis as a key process for life on Earth. Varallyay (2005) recognized soil as the primary food source of the biosphere, the starting point of the food chain. As a spatially (horizontally and vertically) variable, temporally dynamic polydisperse system, soil can simultaneously satisfy—to a certain extent—the ecological requirements (air, water, and nutrient supply) of living organisms, the production of natural vegetation and the production of cultivated crops. He places this ability of soil under the umbrella of one unique property: soil fertility, which varies greatly and has changed considerably depending on natural factors and human activities, such as land use and soil management.

Water supply is a key issue in relation to water content in the soil, because it determines the strength with which water is retained in the soil, which the absorbing strength of plant roots needs to overcome. This process involves the entire root system, with special significance attached to the vast areas of root hairs, illustrated by the fact that for corn at 1 mm root length there are already 422 root hairs (Dubravec, Regula 1995). On the basis of the productive function of soil, which is a factor in maintaining cultural and natural vegetation, agriculture and forestry—two branches that form the backbone of the Croatian economy—supply Croatia with food, feed, fiber, and timber, and are calculated to provide sustainable production of biofuels. Of the total land area of Croatia, 3,212,816 ha or 56.7 % is used for agriculture, and 2,351,270 ha or 41.5 % of the land is used for forestry, which gives 0.74 ha of agricultural and 0.55 ha of forest land per capita (Bašić et al. 2001).

In spite of quite favorable agroecological conditions, yields of the main arable crops in Croatia are below the maximum possible (i.e., the biological potential of the various crop varieties and/or hybrids). The reasons for yield being below maximum (100 % of biological potential) are stated by Mihalić et al. (1981) to be dependent on the soil and these reasons are shown below expressed as percentages:

• Low natural fertility of soils and lack of amelioration	20 %
• Lower quantity or unfavorable distribution of precipitation	16 %
• Unsuitable soil tillage	13 %
• Unsuitable crop variety	16 %
• Unsuitable crop density	14 %
• Crop diseases and pests	10 %
• Other factors	11 %

This assessment was made 30 years ago, but comparing with the recent situation there is no change… or maybe there is, but in the wrong direction, because of decision makers (politicians) favoring the destruction of modern public farms, by fighting for electors, land privatization and fragmentation of already maximally fragmented, practically "atomized" land. Instead of increasing the national wealth by building the infrastructure for land reclamation they allow the destruction of what has already been built! Trees growing in canals for drainage on abandoned land are mature for cutting! (Bašić 2009).

About 95 % of the food that arrives at the table of human society is produced by the soil. The balance in terms of food security for humanity is far from satisfactory; we are still far from any prospective and effective solutions. In spite of these facts, soil is expected to assume the role of providing renewable, economic, socially and environmentally sustainable production of biofuels, which further increases the already (too) high pressures on soil (Tomić and Bašić 2011).

2.2.2 Environment-Regulatory Functions: Soil is a Regulator!

This group of complex functions includes climate regulatory functions, reception of toxic substances (soil is a receptor!), filtering, accumulation (soil is an accumulator!), and transformational activity (soil is a transformer!), between the atmosphere, hydrosphere, and the plant cover (biosphere). In other words, soils protect the environment, and especially humans, through protection of the food chain and underground drinking water reserves.

2.2.2.1 Climate Regulatory Function: Soil is a Source and Sink of Greenhouse Gases!

Soil is a source of emissions of greenhouse gases, primarily CO_2. Namely, soil is the final link in the chain of biotransformation of organic carbon, which results in CO_2 and water as final products—the same compounds that are used in the process of plant photosynthesis. This means that soil is also a sink of CO_2! Layers of soil that are completely saturated with water in the absence of oxygen undergo carbon transformations that end with the formation of methane—CH_4, a more effective greenhouse gas than CO_2. In addition, the transformation of nitrogen fertilizer in the soil results in the emission of NO_x, a quantity of which falls from the atmosphere after electrical discharges during summer rains. As a source of emissions of these compounds in the atmosphere, soil participates in the so-called "greenhouse effect" (Bašić 2009). Although the total content of organic matter—humus in the soil (humosphere)—is low, this negligible (1–3 %) humus is a national treasure of the utmost importance (Toth et al. 2007). Humus regulates chemical and biological processes, it is the food and energy source of soil micro-organisms, or, as stated by Mihalic (1997), it is the "fuel" for the "biological fire" in the soil as a reactor. Its positive impact on soil structure strongly influences the water–air relationship and the hydrothermic conditions of soil, with a great influence on soil fertility (Bašić 2007, 2009). The humus layer of the Earth (humosphere), according to Lal (2000, 2001, 2003), is the third largest "C-pool," and contains 23×10^{14} kg (2,300 giga tones—Gt—billion tons) of carbon dioxide. Globally, the total amount of organic carbon in the humosphere is three times greater than the above-ground biological mass. In the equatorial area it is nearly the same, but in arid-steppe areas this value is ten times higher in soil than in the above-ground mass. Conversely, the "C-pool" of the humosphere—pedosphere is directly associated with the "C-pool" of the biosphere (6×10^{14} kg = 600 Gt) and atmosphere (77×10^{13} kg = 770 Gt). A change in the contents of C in the humosphere of 1×10^{12} kg (1 Gt) corresponds to a change in the concentration of CO_2 in the atmosphere of 0.47 ppm. Thus, an increase of C in the humosphere (which can be achieved by sustainable management of soil) of 1×10^{12} kg (1 Gt) reduces the increase in CO_2 content in the atmosphere by 0.47 ppm. It is estimated that the concentration of CO_2 in the atmosphere by the middle of this century will increase by 50 %. There is no doubt that deforestation, plowing of prairies and steppes, and agricultural expansion in the nineteenth and twentieth centuries caused CO_2 in the atmosphere to increase to a level that may cause climatic changes on Earth. To illustrate this: it is estimated that in the USA the removal of natural vegetation to create arable land, the treatment of these soils and the creation of an agroecosystem caused the loss of between 3 and 5×10^{12} kg (3–5 Gt) of carbon. It has been calculated that about 9 % of the total CO_2 emitted in the EU originates from agriculture. Our estimation is that Croatian soil emitted 723×10^6 kg of CO_2 in 1995 (Mesić et al. 2006), and the total emissions from agriculture and agricultural manufacturers (machinery, fertilizers, pesticide application equipment) amounted to 3.6×10^9 kg of CO_2. Summing the amount of CO_2 emitted by agriculture and other sources, such as forestry, Croatia as a small state represents a small and "innocuous" source, whose "contribution" to the global "greenhouse effect" is a modest one. Conversely, we estimate that the passive form of carbon—durable humus in the soils of Croatia—sinks 225×10^6 kg of carbon annually (Bašić 2007).

2.2.2.2 Soils as Natural Transformers of Various Pollutants: Soil is a Natural Transformer!

Organic substances accumulated in soil are exposed to (micro)biological destruction which ends with the two substances that were at the beginning of the cycle—in photosynthesis, i.e., CO_2 and water. In this way, the soil represents the first and last link in the biological cycling of matter and energy. Thanks to the transformation function of the soil, all postharvest residues in agricultural soils are transformed into the organic matter of soil—humus, including leaves of deciduous trees and conifer needles in forest ecosystems (Martinović 1997).

Similarly, soil transforms all organic pollutants such as PAH, residues of pesticides and petrochemicals into harmless end-products (Photo 2.2).

2.2.2.3 In its Function as a Natural Filter, Soil Preserves Drinking Water!

In its function as a universal cleaner (filter) of rainwater, soil protects underground drinking (potable) water from pollution. The importance of this function is illustrated by the fact that 65 % of the EU population uses drinking water from groundwater (Nestroy 1996). The effect of filtration depends on soil cation exchange properties. The cation exchange process or adsorption of cations is the best known and most important way to protect groundwater from pollution. Specifically, the colloidal complex of soil contains humus and clay. These are the most important negative charged colloids in the soil through which it has the ability to adsorb ions of opposite charge (cations) by forces strong enough to protect against leaching to groundwater and at the same time weak enough for sorption through the roots. The cation exchange capacity (CEC) of the active surface of 1 g of montmorillonite clay minerals is greater than 700 m^2, which is the seat of a negative charge, followed by illite clay minerals, with the smallest CEC capacity belonging to kaolin clay. Cations of plant nutrients but also positively charged radicals and organic compounds, which are initially retained and then subjected to microbial degradation, are adsorbed on this surface.

Biological sorption refers to the binding of environmentally hazardous substances—pollutants in living organisms, particularly plant biomass. In this way soil prevents the leaching of these substances into groundwater and the soil is cleaned of these substances if that plant mass

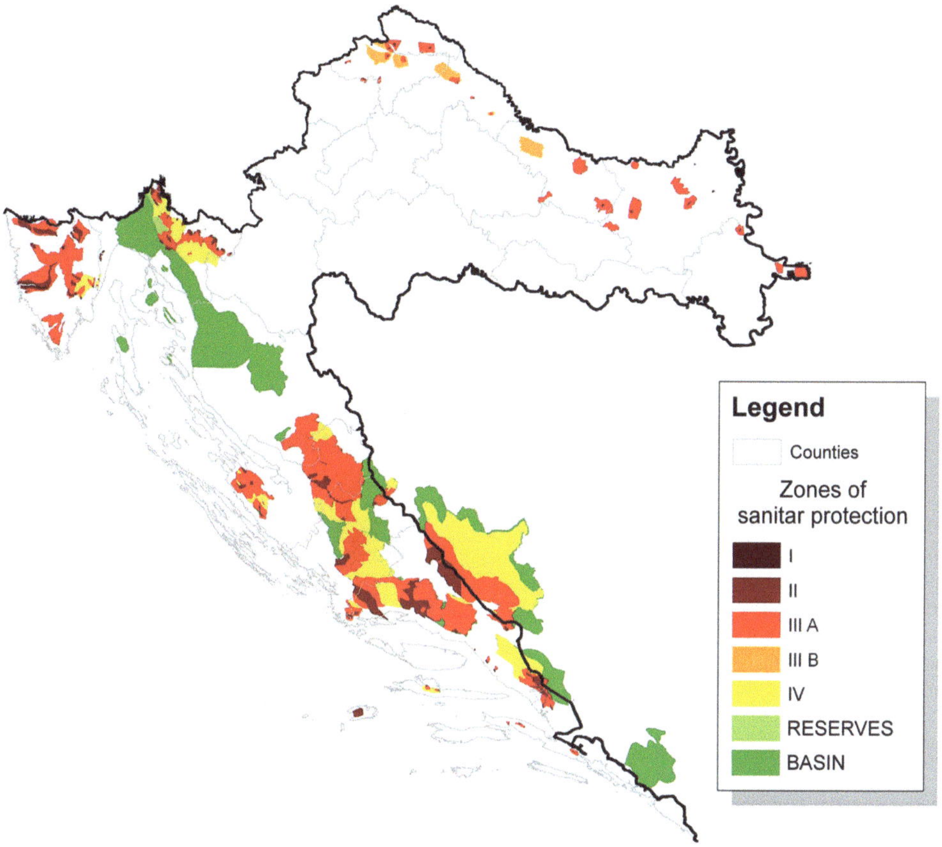

Fig. 2.1 Water protection areas. There are four zones of protection of water; the I one with strongly reduced any activities risky for water eutropication, the IV as most liberal one

is removed from the soil. In this way, by choosing plants that have a selective ability to accept certain pollutants such as heavy metals, soil can be cleaned of contaminants by a phytoextraction–phytoremediation process.

According to soil characteristics, the capabilities and effectiveness of this filtration in Croatian territory differ in the following areas.

- *River basin of the Drava River*. This important area, especially in the western and central parts, is dominated by soils of light texture—gravelly and sandy, very permeable to water with low content of colloids—clay and humus. Pollutants that fall on such soils quickly enter and penetrate into the soil and reach the groundwater. However, these soils also quickly remove pollutants from the soil and clean the water.
- *The soils of the Sava River basin* differ by a significantly higher contents of clay, which accepts and strongly retains pollutants. Because of the poor permeability, pollutants slowly penetrate to groundwater. However, once the soil is defiled by pollutants originating from the upper layers, there is a problem as these soils are slowly polluted but also cleaned slowly. Naturally, in the valley of the river basin, light, sandy soil and gravel is found, as can be seen along the Drava, and also heavy, vertic soils.
- *The karst area* has a special karst hydrology. The quantity of water in numerous rivers and smaller water flows does not correspond to the total surplus of rainwater, which disappears into the highly permeable karst underground. The underground hydrological system is very sensitive and vulnerable; the high-quality drinking water should be preserved from pollution by agriculture—i.e., by nitrates, residues of pesticide, plant nutrients, and heavy metals. Particularly sensitive and vulnerable are soils of karst fields, which are used for intensive farming, and where groundwaters are close to the surface. The spatial distribution of water protection areas is shown in Fig. 2.1.

The most serious polluter of underground drinking water emitted from the soil is nitrate from nitrogen fertilizers, but without nitrogen a high yield cannot be achieved. This problem of world agriculture remains an unresolved issue. In some areas the nitrate from the soil washes away in the post-harvest period after removal of the previous crop, for example in the northern USA states around the Great Lakes. To remediate this problem from sowing to harvest, the use of special "cover crops" has been suggested, whose task is the uptake of the nitrate which remains after removal of the previous crop, to avoid leaching and eutrophication of water in the lakes. Elsewhere, it is wise to consider the limitation of the dose, timing, or form of application of nitrogen fertilizers for the major arable crops. Working as a powerful buffer, the soil system inactivates all substances that rapidly enter its mass or that were released after mineralization of organic

Photo 2.2 Incineration of crop residues is an unacceptable practice. After harvesting of corn a few tons of crop residue remain, from which humus is formed via microbiological processes. The process is desirable because less of the carbon is released as CO_2 and there is greater C sequestration in humus

Photo 2.3 Many bumps on the roots of soybean in the heavy gley soil in the area of Daruvar show the great success of inoculation. These are really a small "biological factory" of nitrogen

matter and thus prevents the stress caused by change in the soil, and shocks to the biosphere—pedoflora and pedofauna. It buffers the acidic components of the soil by using cations such as Na^+, K^+ Ca^{++}, Mg^{++}, etc., and provides resistance to sudden and large changes in soil chemistry. Buffering can be performed by other mechanisms, such as the binding of foreign substances via the colloidal cation exchange complex.

In its biological regulatory function, soil is a habitat and gene reserve for a number of micro- and macro-organisms, or pedoflora and pedofauna; it represents the start and end of biological cycling, is a gene reserve, and is the foundation of biological diversity. The number of living organisms below the surface is many times higher than that on the soil surface.

This is eloquently illustrated by the fact that good, fertile soil in the arable layer contains about 25 t/ha of living organisms, including a number of extremely useful ones, such as symbiotic (genus *Rhizobium)* and nonsymbiotic nitrogen-fixing bacteria (*Azotobacter*, *Clostridium pastorianum*) that use elemental nitrogen from the air and transform it into a plant-available form. This ability of symbiotic fixing bacteria is used in practice for inoculation of legume crops (Photo 2.3). Redžepović (2008), who used selected *Bradyrhizobium japonicum*, which can provide the plant with up to 180 kg nitrogen per hectare, presented outstanding results in this field. The economic value of this work is reflected by the fact that the bacteria on soybean tied 132 kg/ha of nitrogen and reduced the requirements of energy for synthesis of nitrogen fertilizers.

Environmental benefits of soil inoculation using nitrogen-fixing inocula are priceless, because they significantly reduce or completely stop nitrogen from leaching into groundwater and streams (Huić-Babić et al. 2008). The fertile soil has a high biological activity and biodiversity. In addition, fertile soil increases the total number of organisms and their diversity. Biological degradation of soil indicates the degradation of physical and chemical characteristics of soil. Soil's rich life represents a "genetic reservoir," i.e., a vast wealth of genes, from which in the future genetic material will be derived for the different future needs of agriculture. Biotechnology is seeking a way to manage beneficial soil biological processes, moving them in a desired direction. Options in this regard are unpredictable and virtually inexhaustible (Butorac 1999; Bašić and Herceg 2010).

2.2.3 Soil Provides the Space for Settlements and Infrastructure: We are Living, Producing, Transporting, Playing... on Soil!

Soil characteristics have had a major influence on land use, today and in the past. Specifically, the pedosphere provides

Photo 2.4 On the island of Korčula, the settlements were built outside the fertile karst fields. Fertile soil in the field is a guarantee of continued survival

space for agriculture, forestry, expansion of settlements, urban areas, roads, recreational areas, and finally space for waste management. For example, in EU 27 rural area dominates, about 2 % of total area is under buildings and roads, with a range of only 0.5 % in Ireland, 12 % in Hungary, 13 % in Italy, and as much as 14 % of the Netherlands. In Croatia, settlements cover 0.8 % of the total land area, with the least in the mountain region—0.2 %, and the most in the Pannonian Croatia—1.2 %. It is unnecessary to mention that the land area is considered permanently lost for the primary purpose of the soil—the production of organic matter—and treated as an irrevocable—the irreversible loss of soil. It is absurd to demand and expect a complete halt to this process, but it should certainly be brought under effective control. The most efficient tool to achieve this aim is public awareness.

Soil was not taken into account for the selection of road routes, the position of livestock farms, or the creation of settlements—urban residential areas (Photo 2.4). When making decisions on the routes of roads or highways a solution should be the shortest possible route in terms of intersection with valleys or river valleys with fertile soils. For those parts of a route, the use of indoor sewerage systems is reommended, to collect and deposit pollutants. Poor decisions on location can have permanent and far-reaching consequences for the use of these facilities. One good example is certainly among the best in Europe, the route of the Zagreb—Split highway.

2.2.4 The Function of Soil as a Medium for Waste Disposal: All Communal and Industrial Wastes Terminate on/in Soil!

Waste disposal is only one of the spatial functions of soil. The effectiveness of a landfill depends greatly on the soil on which it is located, and the choice of location is very delicate and a highly professional matter. Elimination criteria for selection of the best or least unfavorable locations are well known, and most relate to the closeness of different objects, protected natural areas, water protection areas, cultural monuments, forests of special value, or areas with endemic plants, etc. Regarding the soil, the basic requirement for good accommodation is the location of the landfill in an area that excludes the possibility of emissions of pollutants into the environment, especially into water, but also air or biosphere, i.e., the plant life at or around the landfill. Moreover, to meet that condition, the soil should contain a large amount of colloidal substances, i.e., clay and humus, and in particular those with a high adsorption capacity, such as montmorillonite clay and "mature" humus of high quality and cation exchange capacity. If nature has not fulfilled this requirement, one achieves the same by placing layers of clay on the bottom of the landfill, impermeable to water and perfectly retaining contaminants. The total production of waste in Croatia is 13.2 $\times 10^8$ kg annually, of which 1.2 $\times 10^8$ kg is

Photo 2.5 Nova Gradiška municipal waste landfill situated in a fertile field. Soil under the landfill is not contaminated by the usual pollutants

municipality-based (270 kg per dwelling) and 1×10^8 kg is of hazardous varieties (Kučar-Dragicevic 2006). The EU-27 member states produce annually around 2×10^{12} kg of waste, of which about 4×10^{10} kg is hazardous waste, and 7×10^{11} kg crop residues from agriculture. Counting the total area of the EU-27 and the number of inhabitants, area per capita is 0.86 ha of land area. In terms of area of land, there is less than 4×10^3 kg solid waste, or 2.15 kg/m^2, with a trend of steady growth of around 10 % annually. Part of the waste is incinerated and 67 % goes to landfills. Both procedures cause environmental damage. Landfills are placed on agricultural land, and a discharge from the landfill can be a source of pollution of soil, water, and air, which threatens the health of humans, livestock, and wildlife.

In Croatia crop residue is not treated as waste. Because of the chronic "hunger" of Croatian soils for organic matter—humus, there is a law which forbids the burning of crop residue; instead, the residue is plowed in, using N fertilizers to balance the C:N relation, to allow microbiological transformation into humus of high quality.

From the standpoint of soil protection, the best procedures are to avoid entirely the risk of soil pollution by harmful substances, primarily heavy metals, PAHs, dioxins, pesticide residues, radionuclides, and mineral hydrocarbons (Photo 2.5). For this reason, it is not permitted to allow any substance to enter the soil, including compost, without analyzing the content of harmful substances.

2.2.5 Soil is a Significant Gene Reservoir: A Larger Biomass Lives in Croatian Soils than in the Above-Ground Biomass: Soil Diversity = Biodiversity!

Soil is a habitat and significant gene reservoir for the biosphere and an important medium of biodiversity. A considerable heterogeneity of organisms live in or on the soil or are closely related to the heterogeneity of the pedosphere. This function has particular significance in the stabilization and conservation of Croatian biodiversity, which is very high, taking into account such data as the fact that about 2,250 plant species live on Velebit mountain.

2.2.6 Soil in the Function of Landscape Shaping: Soil is the Base of Croatian Natural Beauty and an Emotional Foundation of Patriotism!

Landscape is the "emotional foundation" of patriotism and the feeling that man belongs to an area. A key function in its formation and maintenance is soil, because it determines landscape features and the benefits of options for possible forms of land use. Man has left "impressions," i.e., "messages," on the landscape. Changing the natural vegetation and entering into the space created by agriculture is the "cultural landscape" created through natural conditions,

Photo 2.6 Removal of natural forest vegetation, changing the natural surface and building terraces has led to a new shape of surface—an anthroscape—in settlement Stipančići near Bol on the island of Brač

which enriches the space, making it even more beautiful, affordable, and attractive for rural tourism. Protecting the landscape diversity of each area is inseparable from the protection of soil in the same area. Anthropogenic influence may be so radical (for example surface terracing) that a natural landscape completely changes into an anthroscape (Photo 2.6).

Fighting for survival, our hard-working ancestors, without any mechanization, prepared a "flowing anthroscape" as a mute witness of past living conditions on Croatian islands (Photo 2.7).

2.2.7 Soil as a Source of Raw Materials: Soil Supports Industrial and Other Production!

Soil is an important source of raw materials, especially for the construction industry. Examples include the excavation of stone, soil, or loess for brick, the digging of clay for ceramic crafts and industry, the use of sand and gravel as building materials, bauxite from red soil (terra rossa), or the use of peat as a raw material for the production of substrates for closed spaces (greenhouses, flower pots). Exploitation of these raw materials is closely monitored in terms of soil damage, for example through opencast mining, or by covering fertile soils with these materials. All mining works, including of course oil and gas exploitation, contribute more or less to soil damage. For example, this happens by removing or damaging the surface, fertile soil layer to accommodate the plant for the exploitation of mineral raw materials, by the construction of access roads that change the natural hydrological conditions, or through the burying of pipelines for transportation of oil and gas. It is estimated that approximately 0.05 to 0.1 % of the land area in Europe has been damaged by opencast mining, and this is not a small area. In Croatia, there are 593 localities used in mining exploitation, which cover an area of 4054 ha, or 0.07 % of the total area (Photo 2.8).

The law specifies a process of soil recultivation and remediation. The obligation of the mining industry is to return the soil to the same state as it was in before mining began. Gas exploitation is also an environmentally hazardous industrial process, because mineral hydrocarbons are mixed with various pollutants, like mercury, mercaptans, H_2S, enormous quantities of CO_2, etc., which must be collected and eliminated from the environment (Photo 2.9). The state of the environment—agroecosystem and forest ecosystem—around a gas refinery is constantly monitored.

2.2.8 Conservation: The Archival Function of Soil: Soil is a Conservator of Croatian Natural and Human Heritage!

Soil conserves information on the conditions of soil genesis and geogenic processes and this information is useful for the interpretation of soil evolution. In addition, traces of

Photo 2.7 Numerous walled fields—in the past cultivated arable land or "sailing gardens"—are today pastures totally unused, on an Adriatic island

Photo 2.8 Oil exploitation equipment on arable land of the Sava River valley (*photo* I. Kisić)

plant life, such as pollen and paleontological material, allow the reconstruction of living conditions of a given area in past times. Soil is an information source allowing the reconstruction of human life and its activities as witnessed by the archeological remains covered by soil and protected from devastation and destruction. For the dating of historical events and changes, archeology is based precisely on these residues.

Completing the explanation of soil functions, the goods and services that are provided by soil are presented in Fig. 2.2, as demonstrated by Blum (2004).

Photo 2.9 After years of oil production the ground is covered with infrastructure and when the wells are exhausted, there follows a procedure of recultivation, i.e., returning the soil to its original condition, as seen at the first oil field in Croatia—Gojlo, near Kutina city (photo: I. Kisić)

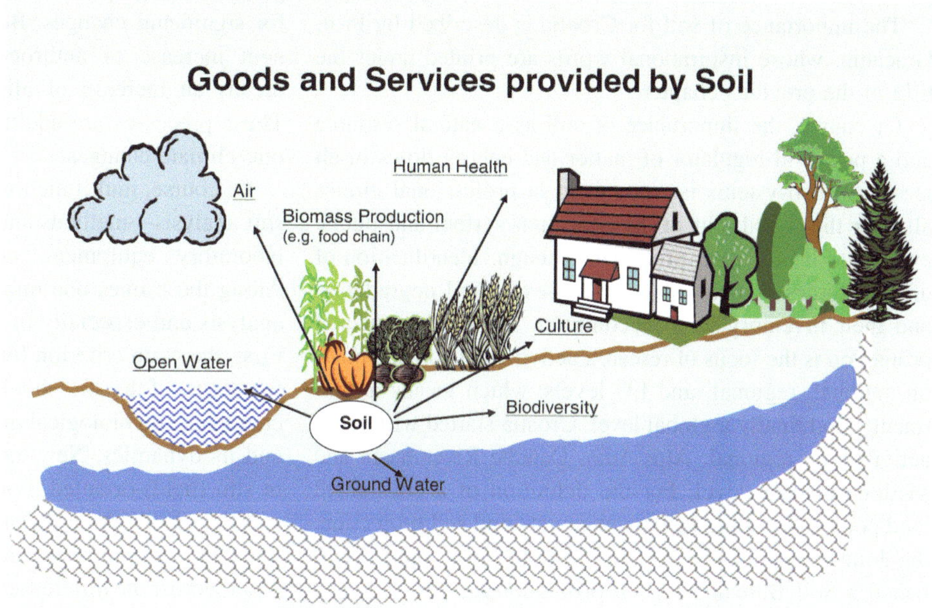

Fig. 2.2 Soil functions (Blum 2004)

2.3 Concept of Multifunctionality (MFCAL) in Land Management

Global trends are fully confirmed by Blum's words. At the crossroads of the millennium—1999—when each message has a "double effect," the functions of agriculture and soil were defined by the International Conference organized by FAO and the Government of the Netherlands in Maastricht, entitled "Multifunctional Character of Agriculture and Land—MFCAL (Mesić et al. 2000). The concept proposes a radical change in the centuries-old criteria for evaluation of soil and the efficiency of agriculture based on quantity and quality of the yield. Instead, after determination of management impacts on other spheres, the focus has been redirected onto "non-food" functions of soil: first, the regulatory function in terms of emissions of "greenhouse gases," the transformation of pollutants, water filtration capacity, protection of water from pollution, etc. It also has an important spatial focus—accommodation of industrial plants, roads, settlements, waste dumps, environmentally, aesthetically and economically acceptable design and protection of landscape, the role of soil in the protection of biological resources as sources of raw materials, and the protection of archeological heritage (Varallyay 2000). The concept is based on the fact that in soil management, depending on the circumstances, any of these roles can become dominant, and everything else may be marginalized, while some are also mutually exclusive.

The problems of soil degradation are caused by the competition that exists between forms of soil use, giving preference to some soil functions, while ignoring or marginalizing others. The new concept of sustainable land use and protection of soil can be defined as the spatial and temporal harmonization of all soil functions in soil and land use, minimizing irreversible ones. Competences for this concept are partly within the scientific realm but more so within the public and decision-maker (politician) spheres.

Of course, not all the described soil functions are carried out simultaneously, and some are mutually exclusive or in competition, which is an important issue for sustainable land management, because one should perhaps favor a second function or marginalize others.

Taking into account all the functions of soil, let us conclude that the soils of Croatia are an excellent natural treasure, a treasure that should be recognized, researched, explored, protected, evaluated and used in accordance with the principles of sustainable development for the prosperity of the nation, and the welfare of its present and future generations.

The importance of soil for Croatia is described by Prof. Gračanin, whose inspirational words are printed under the title of the previous chapter.

Of course, the importance of soil as a natural resource and a powerful regulator of matter and energy flows in all terrestrial ecosystems is recognized in professional circles all over the world, but in the education system and public awareness it is not yet prominent enough. Identification of driving forces and the main processes of soil degradation and their inventory as a precondition for an efficient soil protection is the focus of research activities of soil scientists on national, regional, and EU levels, which means a continental and finally a global level. Croatia started with these activities on regional; Alps-Adia, Danube River basin and Mediterranenean level. For the definition of a considered land policy, including land management and soil protection, the Joint Research Centre (JRC) EU in Ispra, Italy established a Soil Bureau with top professionals, who engaged scientists from all European countries and formed a European Soil Bureau Network—ESBN, with the main task of creation of a European soil policy. Within ESBN, numerous projects have been realized along the lines of knowledge of the soil, and its degradation and protection. Of epochal importance is the first edition *Soil Resources of Europe* (Jones et al. 2005), followed by *The Soil Atlas of Europe* (2005) and the publication on soil degradation (Van-Camp et al. 2004) with six volumes and almost 900 pages.

ESBN prepared the text of a thematic strategy for soil protection, which would be promulgated in the form of a Directive of the EU obligatory for all member states.

There is no doubt that multifunctionality is a new concept, but it already has much influence nowadays and even more on the future of soil science, since this science is at the crossroads of its development, as recognized in the book with the encouraging title: *The Future of Soil Science* (Hartemink 2006). This book introduces the view of selected soil scientists on this sensitive topic. There are two clearly differentiated tendencies. One of them deals with respecting the importance of environment/regulatory ("non-food") functions of soil. This is the case with soil scientists in Europe and developed countries; they expect to see the future of soil science within the environmental sciences. They insist on soil monitoring and "green energy" and more or less accept soil (land) as an essential resource for (environmentally, socially, economically) sustainable production of biofuels. The other tendency is a focus on productive functions of soil (food) and here we see soil science as agriculture and forestry, putting pressure on achieving ever higher yields of growing crops and forestry. Croatian soil scientists are in a transitional position, deep in traditional "food-oriented" ideas but active in research on non-productive ones. The awareness of decision makers and public of the importance of soil is far below that necessary for significant changes. Both tendencies promise a permanent increase of anthropogenic pressure on soil, which means an increase of all processes of land degradation. These processes are additionally stimulated by recent chaotic climate changes.

Of course, multifunctionality of soil needs new, adequate soil analysis—methods and interpretation, using the newest laboratory equipment, computer and nano-technology. Along these lines one might expect a re-orientation of soil analysis and especially of its interpretation. In which way? First, the main criterion for the evaluation of the productive functions of a soil should be the state of all physical, chemical, and biological properties of the rhizosphere of soil and its dynamics. New methods should be more and more in situ (field)-oriented. For example, the redox potential of soil tells more about the state of soil than do single data on pH, CEC, or humus content obtained by the best analysis. In other words: the differences between soil properties obtained from the analysis of a soil sample and the actual suitability of soil in the rhizosphere for plant growing are too big! There is a very wide field of multidisciplinary activities for finding efficient and acceptable solutions to this problem. Correct use of computer technology is of valuable support, but it can lead to a "virtual sphere of soil science". We are witnesses of the consequences of similar tendencies in the world's financial system.

It is true that throughout history, soil (land) was at the root of radical historical changes, from forming the first civilization, Moses coming back to the promised Holy Land, followed by the slave-holder system, feudalism, up to capitalism with the market economy, of which the recent "fruits" are "yellow-colored European fields" of oil rape and/or sunflower for biofuels. By putting so much pressure on soil, mankind is cutting the branch on which it is seated. One can say; there is nothing new on "the blue planet"! It is to be expected that one of the consequences will be that the main occupation of soil science in the next generation will be soil-protection oriented! But contrary to the previous generation, the "maneuvering space" for action is going to be more and more reduced.

In this context we believe that new changes will follow, first in land tenure relationships. It is true that private ownership of land is the basis of recent society and the market economy. But there are already numerous interventions of society in this relationship: from incentives in agriculture to completely "non-market" interventions (similar interventions in the financial system have catastrophic consequences for the global economy). The interventions are on the global level (such as declarations on protection of some sites, protection of biodiversity), the continental level (directives and strategies of EU), and the national level (acts and regulations on protection of protected natural area, water protection areas, good agricultural practices, etc.). Therefore, landowners have to accept all regulations in spite of the fact that it reduces their private property rights. Soil (land) is private, but almost all activities on the land are regulated by society. The mechanism of incentives is compensation, or paying satisfaction of farmers and landowners, but at the same time it disturbs the market relationship on the global market of agricultural products. As we see, because of the vital importance of soil functions it is an unquestionable fact that land management is a very responsible job, which needs efficient use of complex knowledge of soil to begin with. The minimum demand of society is to look for clear and, from the point of view of soil protection, efficient (sustainable) land management. For driving a car it is necessary to have a license, but for land (soil) management it is not! The solution is to change land tenure relationships—land property rights. Just like air and water, as a member of the "ecological triad" soil needs to be a public good and property. Land management for all soil (land) users, including the farming system—crop rotation—has have to be "soil friendly", that is to say, responsible toward society. But products of soil (land)—food, fuel, timber etc.,—are in every case private property and we have competition on the free market. A precondition for a correct land use would be some qualifications, which would include a clear defined minimum of knowledge of soil for land users. In other words, it should be the rule that every square meter of land (soil)—agricultural, forest, urban—has to be under professional supervision of the owner (society). It includes obvious monitoring of soil and changes in land management according to the results.

References

Bašić F (1976) Pedologija, Viša poljoprivredna škola Križevci, udžbenik, str. 231, Križevci

Bašić F (1982) Pedologija, II. preuređeno i dopunjeno izdanje, str. 184, Križevci

Bašić F (1998) Održiva poljoprivreda—Nova "Zelena revolucija", poseban tisak, knjiga: Hrvatska i održivi razvitak, Gospodarstvo–stanje i mogućnosti, Ministarstvo obnove i razvitka, Zagreb, str. 83–103

Bašić F, Bogunović M, Božić M, Husnjak S, Jurić I, Kisić I, Mesić M, Mirošević N, Romić D, Žugec I (2001) Regionalizacija hrvatske poljoprivrede, rukopis, Agronomski fakultet Sveučilišta u Zagrebu, Zagreb, str. 274

Bašić F, Franić R (2003) Croatian agriculture at the crossroads, nature and man in Croatian Agriculture. Croatian food and food processing industry, PRO—TIM, Zagreb, p 82

Bašić F, Kisić I, Mesić M (2003) Alternativna poljoprivreda na vodozaštitnim područjima Zagrebačke županije, studija. Zavod za OPB AFZ, Zagreb

Bašić F (2006) Višenamjensko obilježje kao temelj održivog gospodarenja tlom u svijetlu pristupa Hrvatske EU, zbornik savjetovanja „Sustavsko mišljenje i proces integracije Hrvatske u EU", Zagreb, str. 103–117

Bašić F (2007) Soils of Croatia—status—problems and anticipated solutions. In: Proceedings of the Mediterranean conference on the status of Mediterranean soil resources: actions needed to promote their sustainable use. Tunis, p 87–111

Bašić F (2009) Oštećenja i tehnologije zaštite tala Hrvatske—otvorena pitanja, Zbornik radova znanstvenog skupa Tehnologije zaštite tla i odlaganja otpada, Akademija tehničkih znanosti Hrvatske. Zadar, str. 179–203

Bašić F, Herceg (2010) Temelji uzgoja bilja, udžbenik, Sveučilište u Mostaru, tisak „Grafotisak"—Grude. Mostar, str. 454

Birkas M, Jolankai M, Kisić I, Stipešević B (2008) Soil tillage needs a radical change for sustainability. Agric Conspectus Sci 73(3): 131–136

Blum WEH (2002) The role of soils in sustaining society and the environment: realities and challenges for the 21st century. In: Proceedings of the 17th World congress of soil science: confronting new realities in the 21st century, Keynote lectures. Bangkok, p 67–86

Blum WEH (2004) The future of soil research in Europe: where do we stand and where do we want to go? From TWG research to a European research agenda, European Confederation of Soil Science Societies (ECSSS), BOKU, European Soil Bureau Network, Meeting of the Steering Committee, Plenary and Soil Atlas of Europe editorial board, Ispra

Butorac A (1999) Opća agronomija, Sveučilišni udžbenik. Školska knjiga, Zagreb, str. 648

Durn G (2003) Terra Rossa in the Mediterranean region: parent materials, composition and origin. Geol Croat J Croatian Geol Surv Croatian Geol Soc 56/1 1:83–100, Zagreb

Kučar Dragičević S, Butuči J, Kufrin J, (2006) Zbrinjavanje otpada u Republici Hrvatskoj—postojeće stanje, Arhiv zahigijenu rada i toksikologiju No; 57: str. 263–266, Zagreb

Hartemink AE (2006) The future of soil science. International Union of Soil Sciences, Wageningen, p 165

Huić Babić K, Schauss K, Hai B, Sikora S, Redžepović S, Radl V, Schloter M (2008) Influence of different Sinorhizobium meliloti inocula on abundance of genes involved in nitrogen transformations in the rhizosphere of alfalfa (Medicago sativa L.), Environmental Microbiology, 10(11): 2922–2930

Jones RJA, Houškova, Beata, Bullock P, Montanarella L (2005) Soil resources of Europe II edition, European Commission, ESB, Institute for Environment and Sustainability. Office for Official Publications of the EU, Luxemburg, p 420

Lal R (2000) We can control greenhouse gases and feed the world with proper soil management. J Soil Water Conserv 55:429–432

Lal R (2001) World cropland soils as a source or sink for atmospheric carbon. Adv Agron 71:145–191

Lal R (2003) Testimony concerning soil carbon sequestration by agricultural and forestry land uses to mitigate climate change, to Committee on environment and public works, Washington

Martinović J (1997) Tloznanstvo u zaštiti okoliša: priručnik za inženjere, Državna uprava za zaštitu okoliša, Zagreb, str. 193–221

Mesić M, Bašić F, Kisić I, Bićanić V (2000) Višeznačna uloga poljoprivrede i gospodarenja tlom u Hrvatskoj, XXXVI znanstveni skup Hrvatskih agronoma, str. str. 16-22. Opatija

Mesić M, Fijan-Parlov S, Bašić F, Kisić I, Turšić I, Zgorelec Ž, Sajko K, Vuković I (2006) Agricultural emissions of greenhouse gases in Croatia. In: Abstracts of the State of the Science. Workshop on agricultural air quality. Washington, pp 910–913

Mihalić V, Butorac A, Gotlin J, Čižek J (1981) Agroekološki potencijali primarne biljne proizvodnje u SR Hrvatskoj, Bilten Poljodobra, 29 5, str. 6–11

Mihalić V, Bašić F (1997) Temelji bilinogojstva: udžbenik za srednje poljoprivredne škole, 1. izd., Školska knjiga, Zagreb, str 215

Montanarella L (2007) The EU thematic strategy for soil protection and its implications in the Mediterranean. In: Zdrulli P, Liuzzi GT (eds) Proceedings of the international scientific conference within the Medcoastland project; Status pf Mediterranean soil resources; Actions needed to support their sustainable use, Tunis, pp 15–30

Nestroy O (1996) Poljoprivreda i zaštita voda, Knjiga priopćenja znanstvenog skupa Poljoprivreda i gospodarenje vodama, 13–19, Bizovačke toplice

Redžepović S, Čolo J, Blažinkov M, Sikora S, Pecina M, Duraković L (2008) Utjecaj biostimulatora rasta i fungicida za tretiranje sjemena soje na učinkovitost simbiozne fiksacije dušika. Sjemenarstvo 24:169–176

Resulović H, Čustović H, Čengić I (2008) Sistematika tla/zemljišta, Univerzitet u Sarajevu, Sarajevo, str. 234

Tomić F, Bašić F (2011) Zemljišna politika u službi razvoja Hrvatske, Uređenje zemljišta—preduvjet napretka i konkurentnosti hrvatske poljoprivrede na europskom gospodarskom prostoru, rukopis, publiciran u izdanju HGK; Hrvatska—društvo i država blagostanja, Zagreb, str. 165–194

Toth G, Stolbovoy V, Montanarella L (2007) Soil quality and sustainability evaluation, an integrated approach to support soil-related policies of the EU—a JRC position paper. Institute for Environment and Sustainability. Office for Official Publications of the EU, Luxemburg, p 40

Urushadze TF (2002) Soils in space and time: realities and challenge for the 21st century. In: Proceedings of the 17th World congress of soil science: confronting new realities in the 21st century. Keynote lectures. Bangkok, p 3–23

Van-Camp L, Bujarrabal B, Gentile AR, Jones RJA, Montanarella L, Olazabal C, Selvaradjou S (2004) Reports of the technical working groups established under the thematic strategy for soil protection: vol I—Introduction and Executing Summary, p 126, vol II—Erosion, pp 127–309, vol III—Organic matter, p 31–496, vol IV Contamination and Land Management, pp 497–621, vol V Monitoring, p 653–718, Vol VI Research, Sealing and cross-cutting issues, pp 719–872. Office for Official Publications of the EU. Luxemborgh, pp 94

Varallyay G (2000) Soil quality in relation to the concepts of multifunctionality and sustainable development. In: Proceedings of the symposium on soil quality, sustainable agriculture and environmental security in Central and Eastern Europe, NATO Science Series, vol 69, pp 17–35

Varallyay G (2005) Role of soil multifunctionality in future sustainable agricultural development, environmental management; contribution to solution. University of Zagreb, Zagreb, p 29–39

Zgorelec Ž, Bašić F, Jurišić A, Kisić I, Mesić M, Šestak I, Bilandžija D (2012) Kemija u istraživanju tla, IX. susret mladih kemijskih inženjera, Knjiga sažetaka, 39–39, Zagreb

Conferences and other sources (2001) A framework for indicators for the economic and social dimension of sustainable agriculture and rural development. European Commission Agriculture Directorate-General, 5 Feb 2001

ARGE ALP (2000) Hintergrundwerte für anorganische Stoffe in Böden. Arbeitsgemeinschaft Alpenländer Arbeitsgemeinschaft Alpen-Adria

Communication from the Commission to the Council and the European Parliament (2000) Indicators for the Integration of Environmental Concerns into the Common Agricultural Policy. Commission of the European Communities Cultivating our future, Background papers, FAO/Netherlands

Quo vadis agricultura, HRSG (1997) Absolentenverband der diplomingenieure fur Landwirtschaft, BOKU, Wien, s. 119

Soil Atlas of Europe (2005) European Soil Bureau Network. Office for Official Publications of the EU. Luxemborgh, p 128

Towards environmental pressure indicators for the EU, European Commission and Eurostat (1999) Theme 8-Environment and energy

European Soil Portal http://eusoils.jrc.ec.europa.eu/

Agricultural Regions of Croatia

3

Man... despite his artistic pretensions and many accomplishments, owes his existence to a thin layer of topsoil... and the fact that it rains... .

Old Chinese proverb—from *Soil Atlas of Europe* (European Commission 2005).

As a spatial unit for describing natural characteristics, factors influencing soil genesis, and spatial distribution of soils, we take the most simple but understandable agricultural subregion.

After becoming an autonomous state, an important task for Croatian agroecology was regionalization of the agrosphere, respecting modern tendencies, especially sustainable land management. Based on these facts, within a special project supported by the Agricultural Research Council (ARC) of the Ministry of Agriculture, Forestry and Water Management of Croatia during the period 2000–2003, we undertook the regionalization of agriculture (Bašić et al. 2001, 2007). Agricultural regions are agroecological—spatial units of Croatia's agrosphere, each with its own agroecological conditions. Regions are in turn divided into subregions: smaller units that are distinguished by their specific climate, soils, or relief. Thanks to their specific characteristics, each region provides different conditions for crop growth, differences in farming systems, structures, and types of farm and farm management as well as agricultural techniques in plant and animal growth.

As a territorially small country, Croatia is "a small-sized Europe"—the point of contact of three main European climatc types: continental, mountainous, and Mediterranean. Geological (petrological) and climatic properties and natural plant cover are also very heterogeneous. Thanks to all these influences and the heterogeneity of relief, Croatia has a range of different soil types (Bogunović 1992; Bogunović et al. 1997), with different properties, which ultimately means very wide possibilities for plant growing. As the consequence of the these natural diversities in Croatia, there are three agricultural regions: a Pannonian region, a Mountainous region, and an Adriatic region.

According to the finer division of this diversity, the primary soil type regions were divided into subregions, as shown in Fig. 3.1.

3.1 Pannonian Region

Covering 46.2 % of the whole territory, the Pannonian region is the most important and the largest agricultural region of Croatia with highly developed intensive arable farming and high yields of all crops. It is the western part of the Pannonian plain, formed by sediments of the bottom of the Pannonian Sea. After the disappearance of the sea in the Pleistocene, the whole plain was covered by loess, an eolian sediment. Because in the Quaternary period, including the recent past, the annual quantity of precipitation (rain and snow) increased from east to west , there was a degradation of the highly favorable properties of calacareous loess—leaching, acidification, removal and compaction by water erosion and by glacial processes in the western part. So, in the west, on the surface in Hrvatsko Zagorje (P-4 subregion) are tertiary sediments formed on the bottom of the Pannonian Sea—marl, sand and on the bottom of the hills Pleistocene loams. Therefore, in the P-1 subregion typical calcareous loess generally dominates, in P-2 leached loess, in P-3 Pleistocene loams, but in the P-4 subregion a mix of tertiary sediments—marl, sand, clay, etc. Of course, soil types follow the same logic; we can find all members of the evolutionary series of soils on loess (from east to west); chernozem—cambisol eutric—luvisol—pseudogley. The structure of farm and farming systems changes in the same direction; arable farming—arable farming with animal breeding—animal breeding—mixed farming. In addition, crop rotation illustrates the same changes: two-field (wheat–maize) or three-field crop rotation (wheat–sugar beat–maize) follows increasing of fodder crops including legumes with crop rotation: wheat–soybean–maize; or maize–wheat–maize; maize–oil rape–wheat–maize; wheat–red clover–red clover–maize, etc (Butorac 1999).

F. Bašić, *The Soils of Croatia*, World Soils Book Series,
DOI: 10.1007/978-94-007-5815-5_3, © Springer Science+Business Media Dordrecht 2013

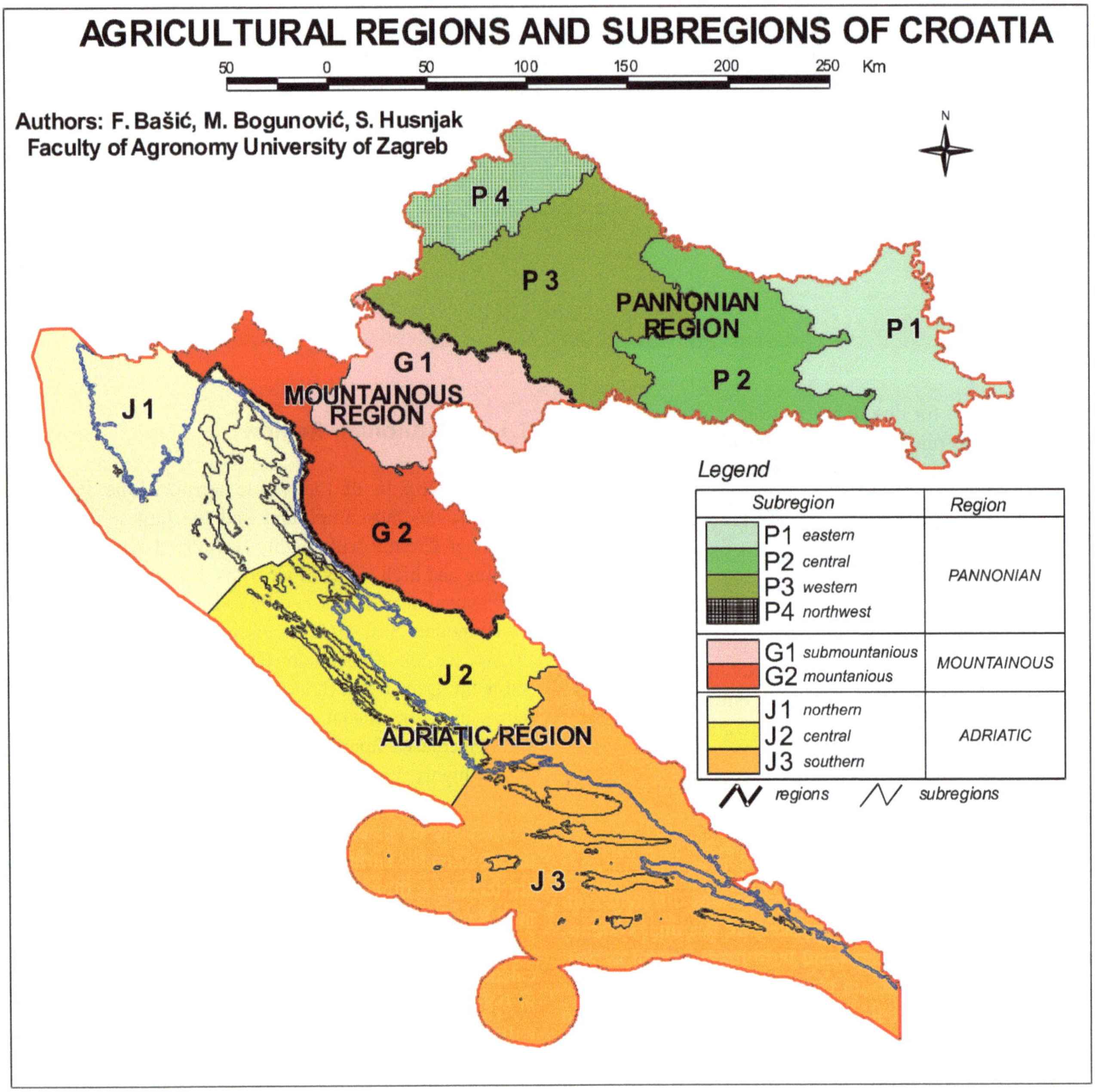

Fig. 3.1 Agricultural regions of Croatia

In the P-1 and P-2 subregions during "the age of the horse" the practice was completely sustainable, the best crop rotation for sustainable land management to date, the so-called "Slavonian three-field" setup; maize fodder (for horses and other animals)—wheat (for bread)—sweet vetch (fodder for horses and other animals). Use of farmyard manure of high quality was regular, the relationship between the three main groups of crops (grain crops of high density, legume and row crops) was optimal from the point of view of pedo-hygiene. All was in balance! Today we can only dream of such or a similar solution: there are no consumers for fodder crops, animal breeding produces liquid manure with problems of disposal…. Our people, say nostalgically: … the good old days!

3.1.1 Eastern Pannonian Subregion (P-1)

This subregion, known as the breadbasket of Croatia, comprises the easternmost part of the country—western Syrmium, Baranja, and the eastern part of Slavonia. The main economic and urban center is Osijek (old Roman

Map 3.1 P-1 subregion

Mursa), followed by Vinkovci (old Roman *Cibalia*) and Djakovo. Here, on the right bank of the Danube River lies the easternmost old town—Ilok (old Roman *Cuccium*) as well as Vukovar (old Roman *Cornacum*)—the martyr town, the hero and the victim of aggression in the last war. The right bank of the Danube was populated from Neolithic times. Near Vukovar is located the famous archeological foundation of the Vučedol civilization, as the "European answer" to Mesopotamia (one Map 3.1).

The subregion covers an area of 605,492 ha, or 10.7 % of Croatia's territory, including 13.5 % of its total agricultural land. According to the census of 2001, the population of the subregion is 491,860 (81 inhabitants/km^2).

In terms of natural vegetation, which has been completely removed, the subregion may be classified as forest-steppe, i.e., it straddles both the forests near riverbeds and the steppes on higher positions consisting of calcareous loess and typical eolian geomorphology. An important

Photo 3.1 Corn and winter wheat in an intensive arable farming area—the dominant crops in this subregion

characteristic of the climate is a chronic deficit of precipitation, so that the main priority in land management, as generally in any dry farming system, is to accumulate snow/rain water during the wet period (autumn–winter) and conserve it for the dry summer. The annual water deficiency in Ilok ranges on average between 1,250 and 3,360 m^3/ha. (Photo 3.1)

The hard-working population of Syrmium and Baranja has a tradition of digging cellars in loess, where they keep their wine to age at practically the optimal temperature. In the past, loess was used for the construction of typical local adobe houses, with walls built by hard pressing of loess.

The climate and typical loess enables formation of very fertile types of soil, primarily chernozem, than cambisol eutric and regosol on loess. These soils provide almost optimal conditions for the cultivation of all agricultural crops, vines, fruits, and vegetables. Vine-growing has a long tradition, starting in the third century A.D., when the army of the Roman emperor Probius brought it there. Vineyards survived the Turkish conquest and were vastly expanded on the renowned estates of Eugene of Savoy, Odeschalchi and Eltz and of the modern companies Agrokor—Belje and VUPIK—Vukovar farm.

The territory of Baranja and western Syrmium are typical examples of Croatia's absurd attitude vis-à-vis irrigation. Namelly, in spite of needs, there is practically no irrigation in this subregion with the most fertile soils in spite is traversed by the biggest river—the Danube with fair quality of water. Great efforts are being made to ensure investment in irrigation infrastructure in that subregion in a manner that would guarantee economic revival and stable development.

The P-1 subregion comprises four internationally renowned vine-growing districts: Syrmium, Baranja, Erdut, and Djakovo. Ilok Traminer—The Prince of Principovac (a famous locality of the Odeschalchi family with the Traminer variety of vine) is the most famous wine in the subregion (Mirošević et al. 2010; Bašić and Husnjak 2010).

3.1.2 Central Pannonian Subregion (P-2)

This subregion stretches through the picturesque areas of western Slavonia, Podravina, Bilogora, and central Posavina. The main economic and urban centers are Slavonski Brod (old Roman *Marsonia*) and Požega (old Roman fortification *Incerum*). The subregion covers an area of 642,073 ha. According to the census of 2001, the population is 382,360 inhabitants—187,300 (49 %) in rural and 195,060 (51 %) in urban settlements. Hydromorphic soils are prevalent—most widespread are pseudogley, hypogley, luvisol on loess, followed by cambisol eutric, leptosol on marl, fluvisol, etc, (Map 3.2).

Intensive arable farming is most prevalent, especially in the flat eastern part of the subregion. The hills and slopes provide good conditions for vineyards and all kinds of continental fruit. This area also comprises "*Valis aurea*—The Golden Valley" and its famous Požega—Kutjevo vine-growing area. The further west one goes in this subregion, the greater the number of cattle farms and fishponds, which receive water from natural water flows. This is the reason why it is so important to protect waters from all sources of pollution. Through erosion, sediment and soil material brings significant quantities of different agrochemicals—nitrate and other nutrients, pesticides, etc. into the water flows and reservoirs—fishponds.

This subregion shows a strong need for land reclamation—Irrigation and drainage including of agroamelioration (Žugec et al. 1987). According to our assessment, pipe

Map 3.2 P-2 subregion

drainage should be implemented in an area of 200,000 ha of hydromorphic soils of this subregion. There is a strong need for irrigation; conditions for it are very good and so would be the effects of such a practice. All the soils in this subregion, i.e., 303,318 ha, are suitable for irrigation, including 88,731 ha (luvisol on loess, cambisol eutric, leptosol on marl, fluvisols), where it can be done without any restrictions, whereas in other areas the soil must first be prepared by drainage of surplus water.

This subregion includes the vine-growing districts Slavonski Brod, Požega, Pakrac, Feričanci-Orahovica, and Virovitica-Slatina. All the vineyards in this subregion are located on the slopes of the central Slavonian mountain range.

3.1.3 Western Pannonian Subregion (P-3)

Located in the westernmost, picturesque part of the Pannonian plain, this subregion covers an area of 1,048,047 ha, of which 402,134 ha is forest, 617,862 ha is agricultural land and 15,296 ha is settlements, roads, and industrial activities (Map 3.3). The area contains a significant share of Croatia's

Map 3.3 P-3 subregion

total population—according to the latest census 1,483,058 inhabitants (33.8 % of the total), with a density of as many as 142 persons per km^2. In dry years, water deficiency during the growing period amounts to 2,740 m^3/ha.

The most common types of soil in the subregion are luvisol on loess, pseudogley, and a very fertile fluvisol, widespread along the Drava River valley; then epigley, hypogley, and amphygley, which needs drainage, followed by cambisol eutric on loess and leptosol on marl and soft limestone.

Investment in land reclamation is the safest and most reliable way to establish stable and reliable growth of all arable crops, vegetables, and plantations of grapes and all continental fruits for direct consumption by the urban population, supply of processing industry, and livestock breeding for supplies of meat, milk, and eggs. Furthermore, it would create new jobs in all those sectors. According to our assessment, this subregion requires drainage of hydromorphic soils over an area of 185,000 ha, but irrigation is viable for vegetable crops grown during the summer planting season on shallow soil and sandy soils. The subregion offers favorable conditions for the cultivation of all kinds of fruit, vegetables, and vines.

The vineyards are located on picturesque, attractive hills, at altitudes between 150 and 400 m. The most renowned vine-growing districts are located at: Plešivica,

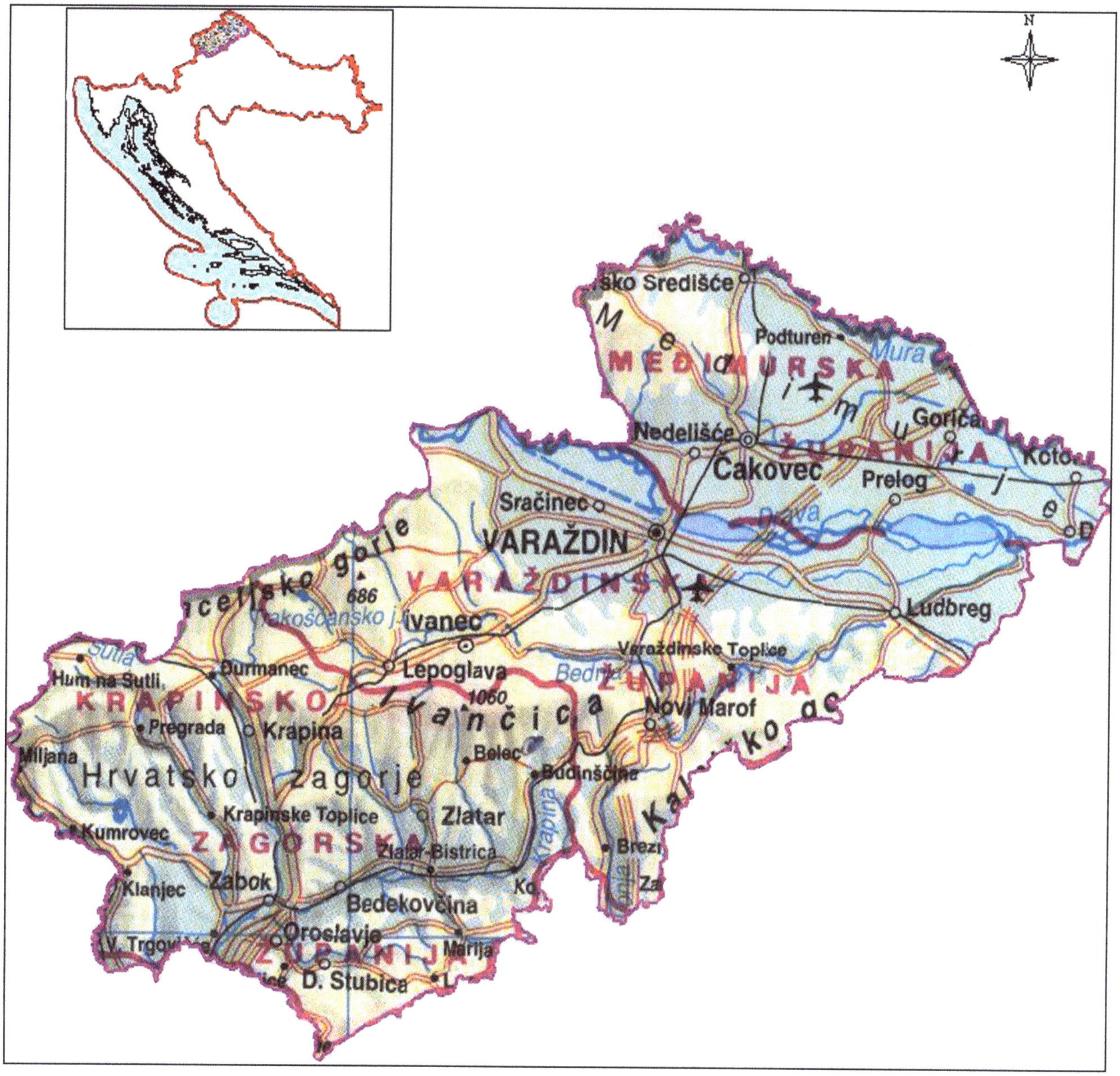

Map 3.4 P-4 subregion

Vukomeričke gorice, Zagreb, Dugo Selo, Vrbovec, Moslavina, Kalnik, Koprivnica, Bjelovar, and Daruvar. The common characteristic of all those vineyards is that they all grow old, autochthonous varieties of grapes of high quality.

3.1.4 North-West Pannonian Subregion (P-4)

This subregion occupies western and north-western Croatia, i.e., the areas of Zagorje, Varaždin, and Medjimurje (Map 3.4). It covers an area of 321,819 ha, with a population of 441,961, as many as 137 per km^2, 55.4 % of whom live in rural areas but work in urban ones.

The subregion features a wide range of heterogeneity; geolithological, geomorphological, and phytocenological, as a result of different picturesque landscapes and a traditionally large population density, which causes high fragmentation of agricultural land and use of marginal land in agriculture.

Because of very different parent materials, leading to a very different relief in the subregion, we find very different soil types. Prevalent are hypogley, amphygley, and semigley, widespread in numerous valleys that have been formed by rivers, rivulets, streams, and brooks, followed by leptosol on marl, pseudogley, and humofluvisols. Most important for agriculture, especially orchards and vineyards, are regosol, anthropogenic soils, and leptosol as undeveloped soils on marl. The common characteristic is intensive water erosion, mostly due to limited permeability of the soils and intense runoff of rainwater.

Characteristics of the climate also contribute to the erosion, but the key factor is anthropogenic influence—the high percentage of perennial plants—vineyards and orchards—with rows oriented up and down the slope, from top to bottom. Erosion is also a consequence of the high frequency of row crops, especially maize, which are tilled up and down the slope. The eroded soil material washed away through erosion causes eutrophication and pollution of water flows and underground water, especially by nitrate.

In this subregion, there are three vine-growing districts: Medjimurje, Varaždin, and Hrvatsko Zagorje. The varieties of grapes are different from vineyard to vineyard. The most prevalent grapes are old and autochthonous, but there are also some new ones. The vineyards of Medjimurje are undoubtedly amongst the best. The prevalent varieties of grape are white: Riesling, Moslavac, Pinot, Chardonnay, Traminer, Sauvignon, and Yellow Muscat.

3.2 Mountainous Region

Covering 24.5 % of the territory of Croatia, the mountainous region is the smallest and minimally populated of the regions in spite of favorable life conditions and good transport connections with Pannonian and Adriatic Croatia.

3.2.1 The Perimountainous Subregion (G-1)

The perimountainous subregion covers 569,403 ha: parts of Lika, Banovina, and Kordun with 169,921 inhabitants or 30 per km^2 only. Drastic depopulation is one of the consequences of the last war.

The subregion rises gradually from the Pannonian plain towards the massif of the Dinara mountain range (Map 3.5), with two historically important, impressive, and scenic mountains: Petrova Gora and Zrinska Gora. Their extension south of the towns of Karlovac and Duga Resa is a large karst plateau with very prominent and developed karst phenomena. The specific landscape feature is the large number of karst holes (vrtača), known in the geological literature as "pock-pitted karst." Its other characteristic is that it is covered with deep layers of well-developed soil—acrisol (two layers).

The average annual precipitation in Karlovac is 1,158 mm and in Ogulin as much as 1,610 mm. Water balance in Karlovac and Ogulin shows an annual water surplus of 470–970 mm. In dry years water deficiency ranges between 52 mm and 135 mm, which is rather high and affects the crops very negatively, especially on the shallow soils on limestone.

The agroecological conditions of this subregion offer very favorable conditions for the growth of all arable crops, and especially fodder crops.

Local cattle breeding has a long tradition and all economic measures for development of cattle breeding in this subregion are of high priority.

3.2.2 Mountainous Subregion (G-2)

This subregion is a prominently forest one, with 61 % of the area forested, and 38.3 % agricultural land. From the point of view of agriculture this is the most backward and least-populated area of Croatia, with ten persons per km^2 only, practically according to all citeria a semi-desert. However, this subregion features a gem of nature—the National Park Plitvička jezera (Map 3.6).

Rather than water deficiency, this subregion has a surplus, ranging from 830 to 1,357 mm. However, in dry years Gospić suffers water deficiency, most often in July and August. The water balance shows that the total deficiency in a dry year may amount to 140 mm, but there are years when the surplus ranges between 1,100 and 1,940 mm.

Although lithologically rather homogeneous, because of a series of other factors that bear on formation of soil, especially the relief and vegetation, this area is pedologically very heterogeneous. The dominant soil type is kalkocambisol on limestone and dolomite, followed by rendzina, and kalkomelanosol, luvisol, and relict terra rossa.

This mountain subregion lacks good arable land, which covers only 0.2 % (477 ha). At least half of that land, especially in narrow valleys along the rivers and brooks, should be left as habitats for marsh vegetation and bird fauna.

3.3 Adriatic Agricultural Region

The Adriatic agricultural region occupies 29.6 % of the total territory of Croatia, from Istria to Konavosko polje, including all islands. A "weekend population" populates the whole area with very high pressure on a narrow strip near the sea, but it also has a large territory of abandoned land of high fertility in the hinterland.

3.3.1 North Adriatic Subregion (J-1)

This subregion comprises the whole of Istria; from there it stretches along the Velebit mountain range as a narrow swathe of land to Starigrad city near Zadar, and to the islands (Map 3.7). Its area is 452,934 ha; the population is 484,853 inhabitants.

Map 3.5 G-1 subregion

The dominant parent rock in this subregion, especially in Istria, is Mesozoic–Cretaceous limestone, but the most well-known and widespread soil type is terra rossa, a very fertile soil with a typical red color and favorable physical, chemical, and biological properties. The part of Istria with terra rossa is known as "Red Istria." There are all members of the evolutionary series on limestone and dolomites; lithosol kalkomelanosol–kalkocambisol–luvisol as well as rendzina, regosol, rigosol, hortisol, and soil of terraces on flysch (mixed marl and sandstone). The part of Istria with these soil types is called "Gray Istria," because of the gray color of flysch as the parent material and soils.

Irrigation is the most important task, a prerequisite for high efficiency and stability of all farming systems in the subregion. In the light of that, it has been established that under average conditions for cultivation one needs to make available 2,400 m^3 of water per ha. However, in some years the deviation from the average may be quite large. All signs point to the fact that irrigation is the top priority for stable growing of vegetables in this subregion.

Irrigation may be carried out with the available surface and in some places underground water. However, in order to ensure the necessary volume of water for irrigation, it would be advisable to build multipurpose reservoirs of various

Map 3.6 G-2 subregion

sizes. The quality of surface water is mostly satisfactory, except near confluences of rivers and the sea, where it is mixed with salty water. In some littoral areas underground waters along the coast are salty and/or alkalized due to the influx of seawater.

In the wine atlas of wine connoisseurs, this agricultural subregion has an outstanding position. Vineyards cover a significant area of the land of this subregion in Istria and the islands. However, in addition to Istria, the vine is an important plant in the area of Hrvatsko primorje and on the Kvarner islands in the north. All of those vine-growing areas have long traditions and certainly in the area of Istria the most common white grapes are Istrian Malvasia Chardonnay, Pinot, Trebljano Toscano, Muscat of Momjan, and other less common quality grapes. The red grapes present there are Merlot, Cabernet Sauvignon, Cabernet Franc, Teran, Hrvatica, etc.

In the vineyards of the Rijeka and Kvarner region, the island of Krk has white Žlahtina and red Brajdica, the island of Susak red Trojiščina, white Krizol and red Suščan, and smaller areas are under white Pljeskunac and white

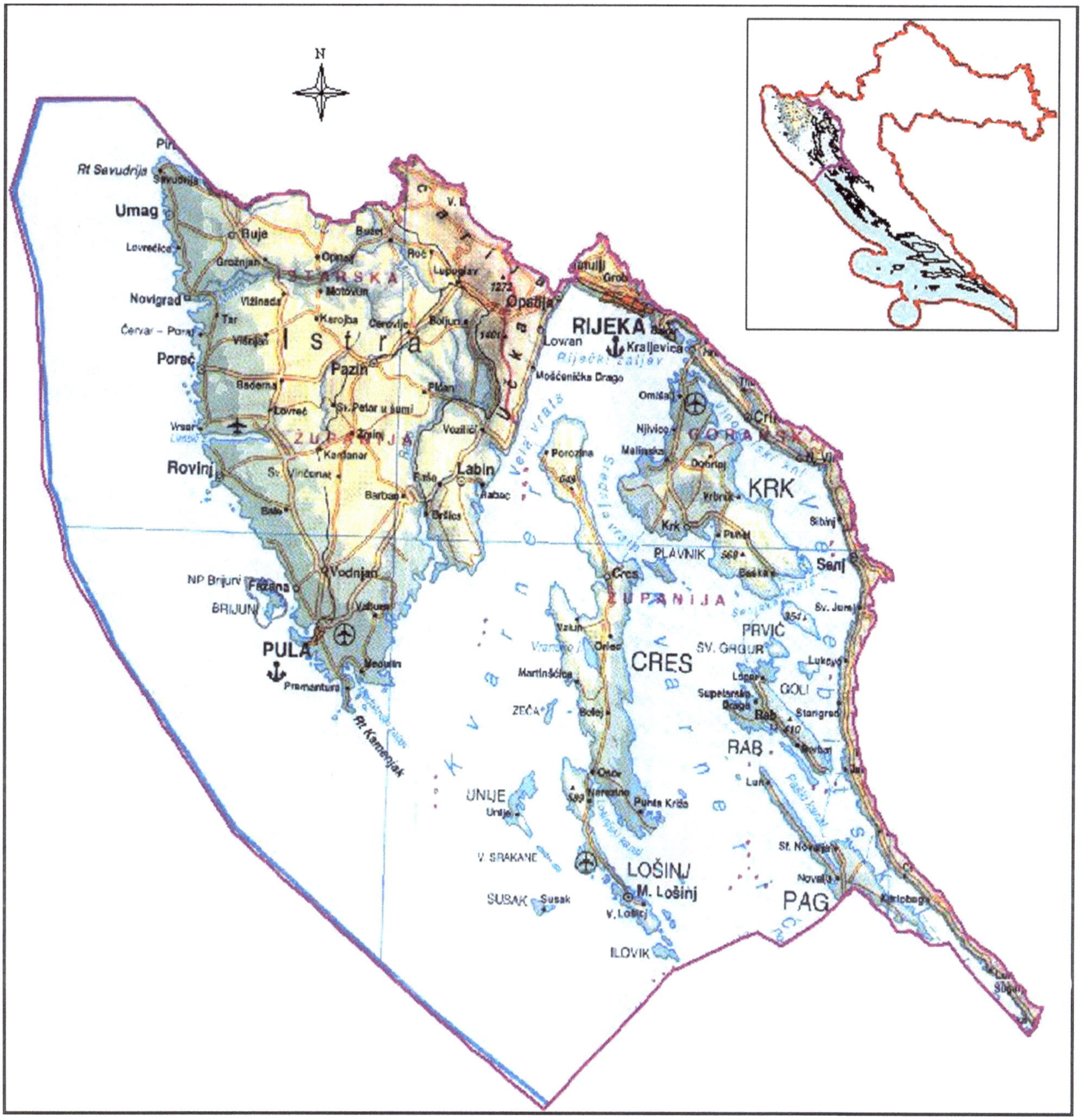

Map 3.7 J-1 subregion

Sauvignon, while other grapes appear only in traces. The most renowned grape on the island of Pag is white Gegić (Mirošević and Veršić 1996).

3.3.2 Central Adriatic Subregion (J-2)

This subregion is in the central part of the littoral area, i.e., the area of Zadar and its hinterlands—Ravni Kotari, as well as the coastal islands and the entire territory of Sibenik County (Map 3.8). The area of the subregion covers 570,946 ha, or 10.1 % of the total of Croatia. The population is 267,171, almost 50 % less than in the territorially smaller northern Adriatic subregion. The ratio of agricultural land per inhabitant in this subregion is 1.53 ha, three times more than in the north Adriatic agricultural subregion, but with stony karst pastures predominant.

The karst plateau is divided by three beautiful watercourses—the Zrmanja, Čikola, and Krka Rivers. The valley of the Krka River has been designated as a National Park and

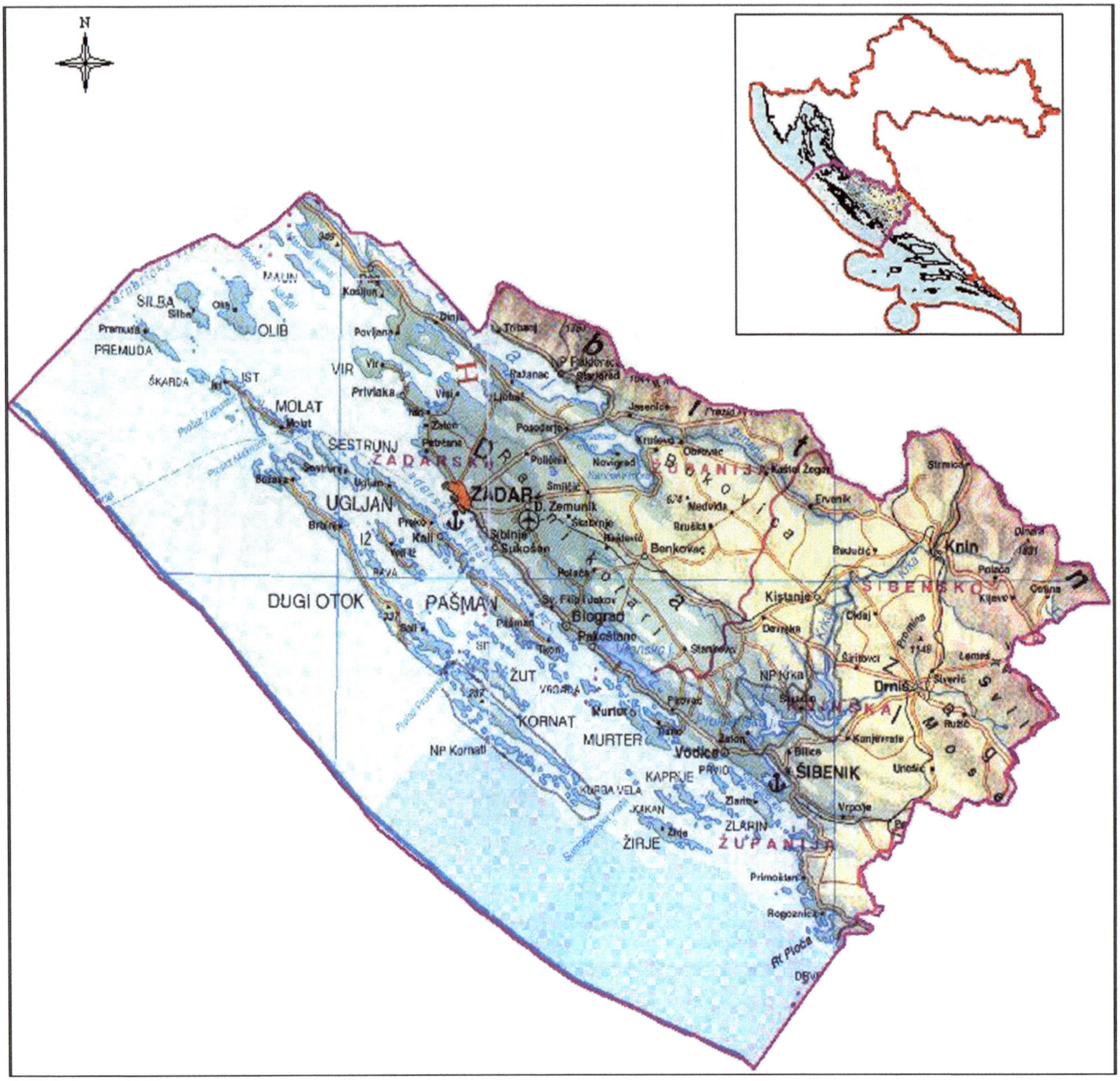

Map 3.8 J-2 subregion

commercial activities there have been strongly restricted. To this subregion belongs our largest freshwater lake—Vransko jezero. In the south of the lake, freshwater mixes with seawater, so that it is only partly and of limited use for irrigation. In the year of writing (2012), an extremely dry year, sea water has penetrated unusually deep into the fresh water.

The subregion consists of a variety of limestone, mostly from the Cretaceous age. The prevalent soil type is kalkocambisol on limestone and dolomite followed by terra rossa, leptosol on hard limestone, and rendzina.

Traditionally, this subregion has had very sophisticated and developed agriculture, especially the growing of vines, fruit, and vegetables, and cattle breeding further back in the hinterlands. In the subregion hydrotechnical land reclamation has already been carried out by the installation of drainage pipe systems on state-owned farms. This was one of the first Croatian regions to introduced such systems. The total land area undergoing hydrotechnical land reclamation is more than 4,000 ha. According to our assessment, in the midterm this subregion will need investment for irrigation over at least 60,000 ha.

In this subregion a stable production of vegetables that are planted or sowed late in the spring or early in the summer is possible only with irrigation.

The needs for irrigation have been analyzed based on the average rate of evapotranspiration measured at the Zadar

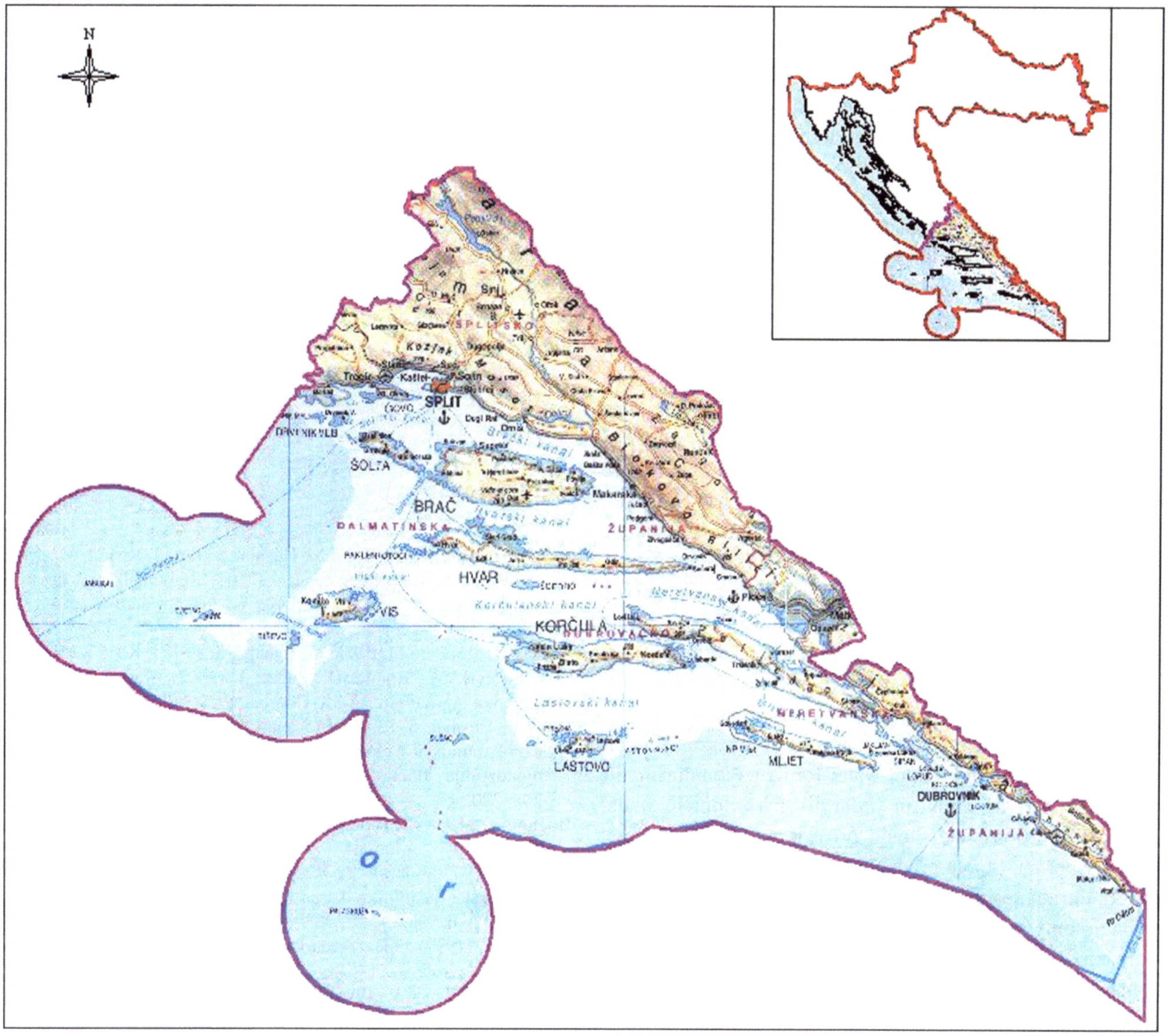

Map 3.9 J-3 subregion

meteorological station as a reference value. For optimal growth and development of crops in the littoral part of the subregion under average climactic conditions, an additional 2,008 m^3 of water are needed for growing melon and 3,013 m^3 for tomato during the vegetation period.

The subregion offers excellent conditions for the growth of high-quality vegetables throughout the year. Conditions are also suitable for the growth of all kinds of Mediterranean fruit, especially olives and figs, but also sour cherries, peaches, and cherries. Vine-growing has a long tradition there. The most important vineyards are those in the areas of Zadar, Benkovac, Šibenik, Drniš, and Knin. The prevalent varieties of grape are Plavina, Lasina, and Gustopjenica (red), and the imported grapes are Carignon, Ugni Blanc, white Grenoche, etc.

3.3.3 South Adriatic Subregion (J-3)

This subregion comprises the rest of the coast down to Dubrovnik and Konavle, i.e., to the state border with Montenegro, including the respective islands (Map 3.9). The area is 631,973 ha, with 578,838 inhabitants. The Neretva River valley is in many ways a special and extraordinary natural complex—a unique marsh habitat, one of the last remaining in Europe. Peaty soils of the valley have been reclaimed by drainage and have been cultivated.

The subregion consists mostly of limestone and dolomites, except for Konavle, where flysch, a mix of marl and sandstone, is the dominant parent material.

The most prevalent terrain is bare limestone karst with all members of the evolutionary series of soils on limestone, except in valleys with a large area under hydromorphic soils; fluvisol, histosol, hypogley and amphygley, both in the Neretva valley and in the karst fields, as well as rendzina, regosol, anthropogenic soils of glass and plastic houses—rigosol and hortisol on flysch.

Given the characteristics of the soils, for stable production and high and stable yields it will be necessary to have hydro- and agrotechnical land reclamation. Over an area of 7,488 ha of hydroameliorated lands the systems have fallen into a state of decay and become dysfunctional and other land has been affected by salinization for a number of complex reasons. Hence, it is necessary to renovate, expand, and repair the systems over an area of at least 2,000 ha. Furthermore, the water level should be regulated in flooded karst fields, the soil in Imotsko and Vrgoračko polje regulated, and pipe drainage should be implemented over a land area that we estimate at 15,000 ha. However, much more important and economically more productive would be irrigation. For that aim we must ensure a sufficient quantity of water, build reservoirs, and obtain appropriate infrastructure and equipment for irrigation of an area that we estimate at 150,000 ha. According to our estimates, the total surface area of land that should be reclaimed by different agromeliorative measures is 115,300 ha.

For optimal growth and development of the crops there is regularly high requirement of water for irrigation during the vegetation period. Thus, in order to grow tomato under average conditions one should provide 4,500 m^3 of water annually per ha. Stable production of vegetables, planted or sowed, during spring and summer months in this subregion is viable only with irrigation.

For irrigation, it would be advisable to build multifunctional facilities for collection of rainwater during the wet season. The quality of most surface waters is satisfactory, except at the confluences of the rivers with the sea. The most important agricultural area is undoubtedly the valley of the Neretva River. Soils in the valley are very high quality, it abounds in water, but because of the influx of seawater, the quality of the water is dubious. Seawater penetrates even further upstream than the town of Metković. Also salinized is the underground water, where the degree of salinity varies.

This subregion comprises the following vine-growing areas: Split, Sinj, Makarska, Imotski, Vrgorac, Neretva, Dubrovnik-Mljet, Pelješac, Korčula, Lastovo, Vis, Hvar, Brač, and Šolta. Each vine-growing area has its own autochthonous cultivars of grapes, but the most widespread are: Babić crni, Plavac mali, Glavinuša, Ninćuša, Okatac, Kadarun, Vranac (red), Trebljan, Mladenka, Maraština, and Debit (white).

Imported cultivars of grapes that were introduced earlier are: Cabernet Sauvignon, Franc crni, and Ugni Blanc. The prevalent grapes in the Imotski vine-growing area are Kujundzusa, Rudežuša, and Vranac red.

References

Bašić F, Bogunović M, Božić M, Husnjak S, Jurić I, Kisić I, Mesić M, Mirošević N, Romić D, Žugec I (2001) Regionalizacija hrvatske poljoprivrede, rukopis, Agronomski fakultet Sveučilišta u Zagrebu, Zagreb, str. 274

Bašić F, Bogunović M, Božić M, Husnjak S, Jurić I, Kisić I, Mesić M, Mirošević M, Romić D, Žugec I (2007) Regionalisation of Croatian Agriculture. Agric Conspectus Sci 72(1):35–38

Bašić F, Husnjak S, (2010) Procjena pogodnosti tala grada Iloka za uzgoj traminca s okvirnim smjernicama gospodarenja tim tlima, monografija Iločki traminac, Golden marketing, Zagreb, str. 203–220

Bogunović M (1992) Pedološka karta Hrvatske, mjerilo 1:2 000 000, Zagreb

Bogunović M, Vidaček Ž, Racz Z, Husnjak S, Sraka M (1997) Namjenska pedološka karta RH i njena uporaba, Agronomski glasnik 5–6, Zagreb, str. 363–399

Butorac A (1999) Opća agronomija. Sveučilišni udžbenik. Školska knjiga, Zagreb, str. 1–648

Mirošević N, Veršić V (1996) Regionalizacija vinogradarskih područja Hrvatske, II međun. simpozij vinogradarstva i vinarstva, Zbornik radova, Zagreb, st. 19–25

Mirošević N et al (2010) Iločki traminac—princ sa Principovca, Ilok Traminer—The Prince of Principovac, monografija, Golden marketing, tehnička knjiga, Zagreb, str. 253

Žugec I, Jurić I, Kovačević V (1987) Neke agromelioracijske mjere kao faktor povećanja plodnosti tla Istočne Hrvatske, Poljoprivredne aktualnosti, 28 (1–2), Zagreb, str. 285–293

Conferences and other sources: Crop water requirement, FAO Irrigation and Drainage, No 27, Rome

Factors of Soil Genesis of Croatia

4

Factors affecting soil genesis include the parent rock, climate (including water), relief, and biological factors—natural vegetation, soil micro- and macroflora and fauna. Some authors distinguish anthropogenic impacts, and others still add time, and vegetation fires, which are particularly frequent in the Adriatic region, to the list of soil genesis factors. Factors of soil genesis are not usually separate and isolated in terms of their influence. The influence of certain soil genesis factors is only in some cases and in particular stages of the evolution seen as pronounced or even dominant, in all other cases they are connected and interdependent. For example, climate influences vegetation, which changes the microclimate, and relief influences both climate and vegetation. By changing such properties, both soil and natural vegetation are changed, and it is therefore clear that in nature the possible combinations of the myriad factors that affect soil genesis thus affect the formation of soil, and all of these factors together result in the great heterogeneity of soil cover in nature (Kišpatić 1877; Gračanin 1951; Bašić 1976, 1982; Resulović and Čustović 2002; Husnjak 2012).

4.1 Parent Substrate

Under the term "parent substrate", we mean a consolidated or nonconsolidated material from which the mineral component of the soil originates. Mineral matter constitutes 80–90 % of the total mass of the soil solid phase. It is understandable therefore that the properties of the parent rock are significantly involved in the properties, flow and direction of soil genesis and evolution. The influence of parent material is particularly important in the early stages of soil genesis and evolution, but later in developed ("mature") soils this influence gradually decreases, and increases the influence of other factors of soil genesis. The chemical and mineralogical composition of the parent substrate determines the properties of the soil that arises from it. The mineral composition of the parent rock also affects the rate of fragmentation and the quality of fragments—its surface, water-holding properties, CEC, mobility, translocation, and distribution down the slope. Released by the weathering of rocks, sandy material provides sandy soils that in the same climatic and under similar other conditions hold less water than soils formed in areas where clay rocks are dominant. Acidic rock is obviously poor in bases and this will give a soil that is too acidic and poor in bases. The "speed" of soil genesis depends on the features of the rocks. A general rule is that soil formation (genesis) on compact rocks is slower than on loose rocks, so that soils formed on this type of rock are shallower.

The degree of the natural supply of plant-available nutrients in soils directly related to the content of biogenic elements in the parent rock. It has special importance for the natural vegetation.

The basic mass of the lithosphere consists of eight elements. They can be divided into the following groups: silicon, "sesquioxide" elements (Fe and Al) and basic elements that are classified as lightly soluble—alkaline (Na and K) and moderately soluble (Ca and Mg). The other elements account for just 1.5 % of the total mass, although their importance is extremely significant, as this fraction contains the largest number of biogenic (essential) elements necessary for life.

The lithological structure of Croatia is very complex, as can be seen in Fig. 4.1, in which 28 mapping units of parent substrates can be seen. The most common parent substrate in a particular mapping unit is listed first, but after this or in brackets are listed much less common substrates within the same mapping unit. The data show that limestone and dolomites are the most abundant lithological units and extend over about 40 % of Croatian territory. This is followed by loess that extends over 19 %, then alluvial (Holocene) loams, clay and sand (ca. 17 %), Pleistocene loams (ca. 7.5 %), marl and limestone (ca. 6 %), sandstones (ca. 2.5 %), flysch (ca. 2.3 %), gneiss, shale, quartz sandstones, and conglomerates (ca. 1.9 %), relict red soil (ca. 1.3 %), and dolomite about 1.1 % of the territory (Husnjak 2012).

F. Bašić, *The Soils of Croatia*, World Soils Book Series,
DOI: 10.1007/978-94-007-5815-5_4, © Springer Science+Business Media Dordrecht 2013

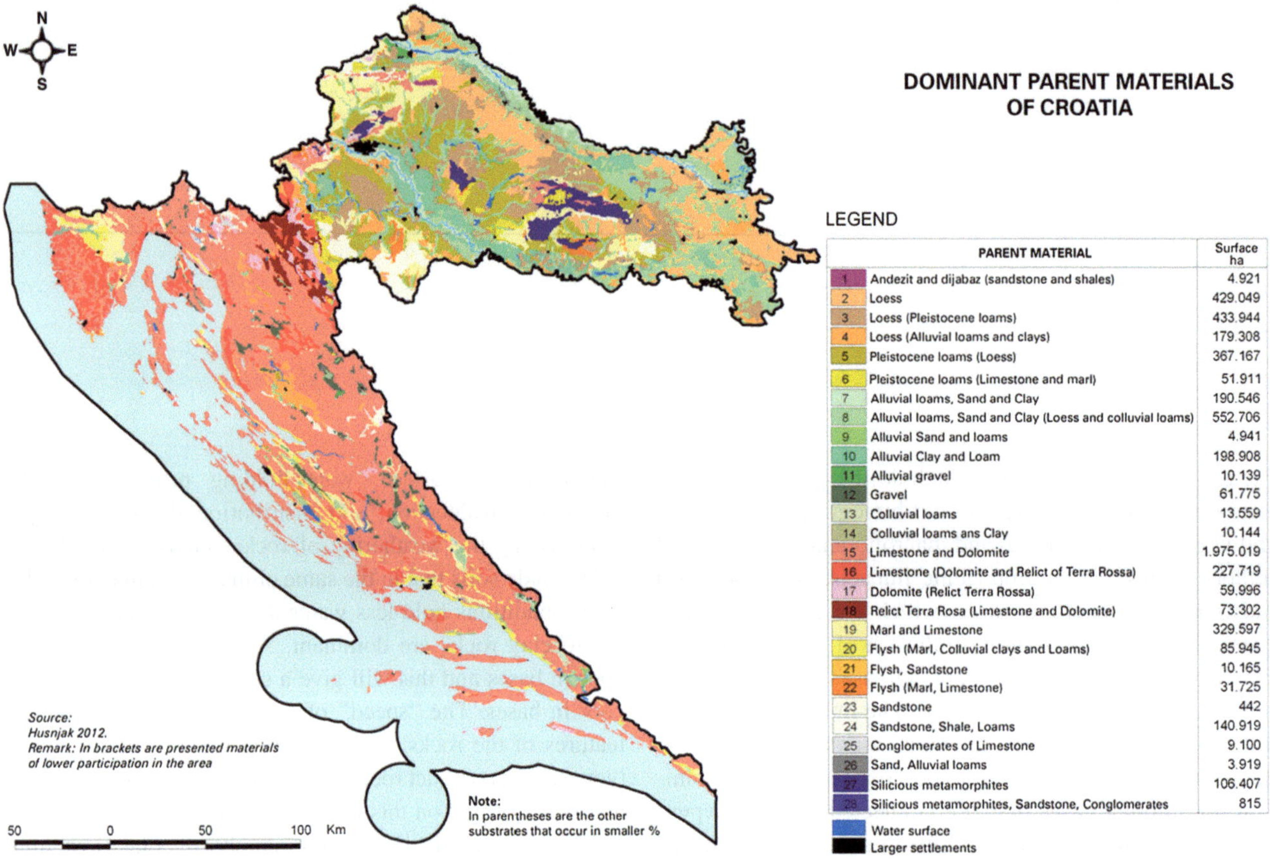

Fig. 4.1 Dominant parent substrates of Croatia

4.1.1 Pannonian Region

The Pannonian agricultural region shows a certain regularity of lithological structure, corresponding to the age of parent materials. The lowest, clearly expressed, are Holocene terraces along the Danube and Sava Rivers and their tributaries in the region. With virtually all the streams are found narrower or wider river or stream valleys, composed of layered alluvial deposits of very different properties, depending on the mode of deposition. There is no particular regularity in deposition in the cross-sections of river valleys, because all the streams, and most especially the Danube, Sava, and Drava, have changed the direction of their riverbed. A common feature is a multilayer structure with fertile soils forming on top, except in cases of high clay content and the occurrence of vertic soils (Photo 4.1). Above the Holocene terrace we find the Pleistocene terrace, which in the entire Pannonian agricultural region is predominantly covered by loess. It is loose, unconsolidated carbonate sediment with a bright, tawny color, and it is of eolian origin. It contains 10–30 % calcium carbonate with a prevailing silt texture, and has a distinctive vertical fracture. Loess emerged in the dry continental climate of one of the ice ages (glacial period) in the Pleistocene. The Pleistocene period is characterized by glacial periods of very low temperature and freezing, with interglacial periods in between with melting of ice. The thaw took place gradually, so that the boundaries of ice moved northwards. In areas of the periglacial zone strong winds blew and brought fine silty material, which covered all other rocks. The calcareous loess in the vicinity of Vukovar is a typical periglacial formation. The glaciers and rivers of the ice age, like a "stone mill," manufactured loess dust that covered today's deposits through secondary deflation. This is shown by the relatively homogeneous grains in the loess. Eolian dust during the foundation of the loess showered the land surface, wetlands, rivers, and other bodies of surface water.

This material, depending on the site of deposition and climatic conditions from the time of deposition until now, has experienced various changes—diagenesis (Galović 2009). So today the Pannonian region is dry, arid, with a typical "real" or carbonate loess (Photo 4.2) in the eastern Pannonian subregion, Syrmium and Baranja, followed to the west by a slightly wetter zone with leached or non-calcareous loess (central and western Slavonia, western Croatia, Bilogora area). In the western and north-western subregions appear the so-called Pleistocene loams that are described by us as sloppy loams (German: Gehangelehm).

Photo 4.1 Deep gravel sediment in the upper part of the Drava river valley near Varaždin

Photo 4.2 Typical calcareous loess with paleosol (*left*); close-up of the paleosol (*right*), eastern Pannonian subregion, city of Ilok

A special type of loess is swampy or marsh loess, eolian material that is deposited in the swamps and waters and is chemically altered—showing an increased content of reduced compounds and gley phenomena in the loess profiles. The most fertile soils in the world are formed on loess. The loess was deposited onto the surfaces locally and formed deposits over 100 m deep, such as in the Vukovar loess plateau. It sedimented in marshes and lakes that covered a significant part of Pannonia, of which Lake Balaton in Hungary is a relict, and formed loess wetlands in which the groundwater level oscillates—in the summer being very low (7–10 m from the surface), and in the autumn–winter period rising to the surface (Škorić et al. 1977; Bogunović et al 1984, 1996; Bašić and Bašić 2007).

The Pleistocene period is characterized by an alternation of freezing during glacial periods and melting during interglacial periods. Soils were formed in the interglacial periods, and were later covered by new glacial loess. Today, on soil cuts they are seen as paleosols—a brown or yellowish layer in the loess. Photo 4.2 shows a particularly well-developed paleosol from the last interglacial (Wurm), in Ilok, western Syrmium.

In some places, such as on the slopes of the hills of Fruška gora in Ilok or Baranja hill in Branjsko brdo, the paleosol is exposed on the surface due to long-term erosion and the upper layer has been mixed with the recent soil by plowing (Kustura et al. 2008). In this case, the morphology of this soil material is similar to recent eutric cambisol, but unlike it contains calcium carbonate.

Loess is a material with very favorable natural properties; it does not prevent the penetration of roots, so roots can penetrate several meters deep, as is visible in Photo 4.3.

From the loess of the eastern Pannonian subregion easily soluble salts are leached, washed out, while the moderately

Photo 4.3 Root of *Robinia pseudoaccacia* penetrating very deep into loess—Šarengrad in eastern Pannonian subregion

soluble salts such as $CaCO_3$ remain after the leaching phase, so that the loess is calcareous from the surface. At several sites, such as in a forest near Tovarnik Spačva, the groundwater, which is saturated by the easily soluble salts, rises to the surface, forming salty soils. Moving westwards from the P-1 to P-4 subregions, annual rainfall increases and leads to leaching of $CaCO_3$ and $MgCO_3$, i.e., moderately soluble salts. Leaching of calcium carbonate takes place gradually, so that in the loess, at the appropriate depth of cemented clay, quartz, and calcite, so-called loess dolls or loess dwarfs (Photo 4.4) form, with their characteristically variable shapes.

In the western Pannonian subregion (P-3) torrential water flows, after the melting of ice in the interglacial periods, caused erosion which transmitted loess to the lower relief positions and foothills of the Middle Slavonian mountains (Dilj, Psunj, Papuk) and Bilogora. Here acidification occurred, and the freezing left very clear traces in the profile that differs from the rust- and gray-colored zone, as shown in Photo 4.2. Unlike typical and leached loess these deposits are compressed and impermeable to rainwater and they are called Pleistocene sloppy loams (German: Gehangelehmen; Photo 4.5).

A Tertiary terrace rises above the Pleistocene and forms the base of the Slavonian Mountains, creating the vivid relief of the Croatian Zagorje and Medjimurje, followed by the central (P-2), western (P-3) and north western Pannonian subregions. The most common substrate on the Tertiary terrace is marl, but on the margin of the Drava river valley and in Bilogora there are Tertiary sands. These sands were originally calcareous but then were exposed to high quantities of precipitation so that $CaCO_3$ was washed out and the sands are acid today.

Marls are sediments of the Miocene rich in calcium carbonate and of different textures: clayey, sandy, or loamy. These sediments were deposited in numerous, shallow lakes that remained after the withdrawal of the Pannonian Sea. Streams brought material into the lakes, which was agitated by currents, so the sediment contains very heterogeneous material. Marl and all soils generated on marls show a great affinity to water erosion. This process is stimulated by the relief situation because these marls are on slopes, as seen in the photograph (Photo 4.6).

These are loose, unconsolidated parent materials very prone to erosion, especially eolian erosion. Before the stabilization of the sands in Podravina, they were mobile ("living sands") and this landscape was called the "Croatian Sahara" (Špoljar et al. 2006). At the top of the Middle Slavonian Mountains are surface outcrops of eruptive or metamorphic, mostly silicate rocks like gneiss, shale, or limestone upon which shallow acid soils covered with trees are formed.

4.1.2 Mountainous Region

The Holocene terrace of the G-1 agricultural subregion consists of alluvial sediments from the Kupa, Glina,

Photo 4.4 Loess dolls or dwarfs

Photo 4.5 Detail of profile of Pleistocene sloppy loam

Mrežnica, Korana, and other rivers. Their common feature is a multilayered structure, the layers of which are heterogeneous. A Pleistocene terrace rises above the Holocene and is much more broken up, developing a flat slope in Zrinska and Petrova gora. The Pleistocene terrace consists of Pleistocene non-calcareous, tight, acid and water-impermeable loams. The Tertiary terrace is very pronounced, and consists of very heterogeneous materials: marl (Photo 4.6), marly limestone and limestone lithothamnium, and there are also heavy clays and Pliocene sands (Photo 4.7). A mountain range rises above the Tertiary sediments with two historically important, impressive, and scenic mountains, the Petrova and Zrinska gora, which consist of metamorphic rocks.

The southern, "inner" part of the premountainous subregion (G-1) is defined as "low or covered karst," so the area consists of limestone and dolomite with all the karst features, but it is covered by a relict terra rossa of Tertiary age (Photo 4.8).

Photo 4.6 Eocene Marl under the Kalnik hills near Križevci (*left*). In a glorious landscape the image shows the mosaic shape of white-colored fresh plowland with glimpses of marl on the surface of the land (*right*)

Photo 4.7 Acid siliceous Pliocene sands at Turanj near Karlovac

Terra rossa covers the whole field, so that one only finds bare limestone or dolomite partially protruding from the red soil material. Terra rossa was covered by a thin layer of loess-like material in the Pleistocene that is still seen in some places on the surface; it seems to be a polygenetic soil, and sometimes it is dispersed so that the red soil is seen on the surface. The extension south of the towns of Karlovac and Duga Resa is a large karst plateau with very prominent and developed karst phenomena. The specific feature of the landscape is a large number of karst holes and karst valleys, known in the geological literature as "pock-pitted karst." Its other characteristic is that it is covered by deep layers of well-developed soil.

From the geomorphological aspect, the submountainous part of the Karlovac hinterland up to the Mala Kapela massif, representing low or covered karst, is clearly distinguished from the mountainous subregion (G-2) of the Velika and Mala Kapela and Gorski Kotar.

The relief of this subregion is strongly dissected, with numerous steep slopes and erosional features. Of great importance for field crop production are the specific geomorphological features—karst fields (Photo 4.9), large depressions in the mountainous zone, as a rule, mountain-encircled valleys, with disappearing streams that flood the fields at a certain time of the year. The fields are filled with limestone and dolomite rock debris, predominantly well-rounded, i.e., gravel. The largest karst fields are Ličko Polje (46,500 ha), Gacko Polje (8,000 ha), Krbavsko Polje (6,700 ha), and Ogulinsko Polje (6,300 ha).

The dominant parent rocks are limestone and dolomites of the Mesozoic era. Their common feature is that they are consolidated rocks and the limestone is of high chemical purity, so that the content of $CaCO_3$ in it ranges from 96.76 to 99 % and it therefore forms part of the mineral soil and there is only 1–4 % of so-called insoluble residue (Škorić et al. 2003).

Photo 4.8 Relict terra rossa as residual soil (paleosol), covered by loess-like material, typical of a polygenetic soil—Bosiljevo near Karlovac

Photo 4.9 Krbavsko polje, a large karst field near Udbina, used as meadows and pasture

The freed insoluble residue was subject to erosion by wind and water, so soils on limestone are shallow; the depth depends on the "underground relief" (Photo 4.10). Deeper soil we find in the topographically lower positions in the cracks of rocks, as shown in the photograph, or in karst holes and fields in large depressions. There appear to be a number of varieties of limestone, massive, platy, thickly layered, and a number of transitional rocks to dolomite: dolomitized limestone and dolomite limestone. Dolomites contain insoluble residue, so soils on dolomite are to an extent deeper. Moreover, their different structure is characterized by coarse-grained and fine-grained dolomite and dolomite breccias.

The series of acid metamorphic rocks are substantially different from the limestone–dolomite rocks. These rocks constitute most of the Gorski Kotar region. These include

Photo 4.10 Limestone with great variations of "underground relief"

the Paleozoic sandstones of Gorski Kotar, the Middle Carboniferous sandstones (Sveti Rok, Štikada), sandstones and shale (the Velebit slopes south of Medak), sandstones (Delnice), shale, clay shale, quartz conglomerates (Crni vrh, Brušani), etc.

These parent materials are poor in bases so they generate a series of acidic soils, whose soil evolution ends with podzol, which occurs sporadically around Zalesina in Gorski Kotar.

4.1.3 Adriatic Region

The Adriatic region from Ćićarija mountain in Istria to Konavle in the hinterland of Dubrovnik belongs to the Dinaric mountain system, which, as shown in Fig. 4.1, consists mainly of limestone and dolomite of Mesozoic age, but is dominated by Cretaceous limestone. Their basic characteristic is the differences in purity or content and the features of the so-called insoluble residue. The content of insoluble residue ranges from 0.5 % to several percents in the Tertiary limestone. The rate of weathering, fragmentation, and dissolving of limestone depends on the features of the limestone, its age, residue content, climatic conditions, and the surface exposed to the forces of weathering. What is more, on fragmented rock, soil will develop quickly and be deeper. Limestone can be massive, hard rock (Photo 4.11) characterized by removal of the insoluble residue via water, wind, or snow and glaciers (Škorić et al. 1987).

Limestone can be brittle after fracturing during tectonic uplift, visible fractures that are often the sites of erosion creating fissures and crevices in the limestone; this gives it a "broken" look, and the insoluble residue can be lost in the cracks, moving deeper and outside the reach of roots or deposited in cracks in the limestone and serving as a substrate for vegetation adapted to such unfavorable conditions (Photo 4.12). Of course, these cracks accumulate all sorts of material that have been brought in by glaciers, water, wind, and gravity in an unselective fashion, thus forming polygenetic soils. The process of the formation of limestone and dolomite as parent materials was time-consuming, and their structure is not homogeneous. Within them one finds securely embedded volcanic ash from active volcanoes at that time. Proposing one theory of the formation of the red soil (terra rossa), Gračanin says that the material is derived from limestone incorporating red volcanic ash with the ash later freed as the limestone is dissolved, which is under recent conditions more or less stable (Durn 2003).

The removal of these rocks to make space for vineyards can lead to vines of high quality.

The wall is erected by annual clearing of the soil to prepare the "living space" for the vine (Photo 4.13). Flaggy limestones with profiles showing different thickness and numbers of flaggy layers are wide spread (Photo 4.14). Some, especially the Miocene limestones, are soft and easily weatherable. A large area of such limestone can be found in the Zadar hinterland, around Benkovac, which is famous as a decorative building material known under the commercial name "Benkovac stone."

Angular nonconsolidated fragments of limestone and dolomite of different dimension are moved by water, snow, and gravity "flows" down slopes and deposit at the foot. If surrounded by walls ("dry walls") for protection against new inflow, such a material can serve as a parent substrate

Photo 4.11 Limestone as a consolidated rock: bedrock on Velebit. There are visible marks of glacial weathering. The products of weathering are redistributed by ice, water, wind, and gravity

Photo 4.12 Hard limestone with fissures and crevices (*left*); insoluble residue accumulates in the cracks in the limestone (*right*). The site is the belvedere of Lake Vrana

Photo 4.13 A vineyard in a dry-walled plot on "naked karst" and shallow stony soil on fragmented limestone in Ravni kotari, the hinterland of Zadar

Photo 4.14 Flaggy limestone with shallow stony soil; Benkovac (*left*). Grinding of this material makes anthropogenic soil—vitisol—for planting vines (*right*)

for soil formation, especially if it contains an amount of insoluble residue of limestone. Such sites on limestone and dolomites are called talus—flowing fragments or "fragment banks."

Without correct protection of such localities by terracing, or by orientation of rows of vines along contours as in Photo 4.15, and/or formation of long plots as in Photo 4.16, the risk of erosion is too high.

Special parent materials are limestone–dolomite breccias, which are angular fragments of limestone and dolomite cemented by $CaCO_3$ in the process of calcification. Similar materials are conglomerates, in which lime material binds gravelly material into a hard rock, which serves as a substrate for soil formation. Conglomerates and breccias usually occur in topographically stable terrain—foothills or in karst fields.

Flysch (from the German: lessen-flow) is a classic sediment which consists of alternating layers of marl and other pelites with layers of sandstone, in which admixtures can occur of angular or gravelly fragments of conglomerate, breccia, and limestone (Photo 4.17). It is widespread in the entire Adriatic region, from Savudrija through to the central Istrian (Buje Pićan) islands, Vinodol valley hinterland of Zadar, Sibenik around Vrpolje, then in Solin and moving on to Omiš and Mimice, and then again in the Konavle area. It was generated by sedimentation of weathered material that composed hills around shallow lakes or seas. The lithologically significant difference that is important for soil genesis is that between the pelitic components (marl) as a loose material, and sandstone as a hard component of the flysch sediment. Depending on the relationship between these components, flysch is divided into normal flysch with approximately equal participation of marl and sandstone, and sandstone and marl flysch, in which one of the components dominates.

The properties of soils generated on flysch depend on which of the components is exposed on the surface to other factors of soil genesis. If this is marl, there is no barrier to

Photo 4.15 Colluvial process of formation of talus as an accumulation of flowing angular fragments of limestone moving to the foot of a limestone slope (*left*). After terracing this material, mixed with insoluble residue, is a good site for a vineyard—(front of) Pelješac (*right*)

Photo 4.16 Vineyard planted on talus—fragments of limestone on the base of steep slopes under Biokovo near Makarska with rows oriented in the direction of the slope due to the risk of water erosion

root penetration for vegetation, which is possible in the sandstone layer. Vegetation cover protects the soil from erosion and allows the soil genesis process of humus accumulation.

All soils on flysch are often affected by water erosion and landslides, which often assumes catastrophic proportions, especially on the flysch of Istria as shown in Photo 4.18.

Damage resulting from soil erosion on flysch is huge at the site of removal, on agricultural and forest land, along roads, canals, river valleys and coastal seawater, which becomes shallower at the mouth of the Mirna river in Istria due to the huge amounts of incoming sediment. Because of constant erosion of flysch the soils are young and undeveloped, and all characteristics, especially fertility and use for agriculture, largely depend on the depth of their pelitic material (marl), which allows rooting. The only effective protection for soils on flysch is soil terracing, using sandstone from the flysch as "building blocks" to build the wall.

Apart from the parent materials described there are others, more of interest as natural phenomena, like loess and

Photo 4.17 Typical flysch with thin and discontinuous sandstone layers—Buzet (western Istria)

Photo 4.18 Erosion on flysch of Central Istria (Buzet)

eolian sands, which occur at several localities in Istria—for example Savudrija, a large surface area forming the entire southern part of the peninsula, to Cape Kamenjak, the southernmost point of Istria. Loess covers the islands Zeča and Unije, but without doubt the most interesting phenomena are reflected in its preservation on the island of Susak (Photo 4.19), which provokes the attention of scientists and naturalists. A complete scientific answer as to the origin and especially the reasons for the preservation of this layer is still lacking, and it is still called the *insula incognita*. The base is limestone covered with a layer of red soil (terra rossa), and loess was deposited on the surface and is preserved on the island today. Along the coast the complete lithostratigraphy is visible. Reed (*Phragmites sp.*) and bramble (*Rubus fruticosus*) were used as efficient protection of the loess against eolian and water erosion. Susak is today a picturesque island of reeds, bramble, and vines.

In addition, eolian sands of Pleistocene age occur at larger and smaller sites around Zadar, with a larger area around Nin.

River and stream valleys in the Adriatic region are composed of very diverse multilayered deposits of alluvial

Photo 4.19 Loess on the island Susak. Water erosion forms typical "loess form", narrow passes cut into the mass of loess (*right*)

Photo 4.20 Neretva river valley, with hydromeliorated peat soils

material. A large area of this material is deposited at the mouth of the Dragonja River, then along the River Mirna, in Čepic polje near Labin, and along the Zrmanja and Krka Rivers. Certainly the most valuable and largest area is on both banks of the valley of the Neretva River in Croatia, especially at the delta mouth into the sea. The locality was once a marshy swamp with lush vegetation that formed a thick layer of peat. In the valley hydromelioration, in particular drainage, is carried out (Photo 4.20). The dominant process in some peat-rich soils is intensive mineralization of organic matter. The network of channels is used for navigation and transport by boat.

4.2 Climate and Hydromorphism as a Factor of Soil Genesis

The average state of the atmospheric climate of Croatia is a factor in soil genesis, and it is extremely variable in time and space. The most important indicators of climate, the total amount, intensity, and annual distribution of precipitation, determine the dominant process weathering of parent rock, leaching and acidification of soil, eluvial–illuvial processes of clay mineral migration, runoff of rainwater, and translocation of soil by water erosion. The thermal regime of the atmosphere and wind influence water balance, evaporation and transpiration, water deficit and/or surplus. All climatic factors affect the genesis of the soil, affecting the type and intensity of the dominant soil genesis processes. Precipitation, heat, and gases generated by the soil primarily cause physical, chemical, and biological weathering, and remodeling and movement of matter in the soil. Wind causes eolian erosion, and enhances evaporation, transpiration, and the occurrence and spread of forest fires, all of which have a significant impact on the genesis and development of soils.

Croatia is geographically situated in the continental part of Europe, which is exposed to the mixed effects of three types of climate: Mediterranean, Mountainous, and Continental. The Dinaric mountain system forms a natural barrier to air currents and the influence of the Mediterranean climate on the continental hinterland. Conversely, the Pannonian plain is to the north and east open to atmospheric currents of cold air masses. Central Croatia has a moderate continental climate, characterized by quite clearly differentiated seasons—spring, summer, autumn, and winter. There is a large annual range of temperature. It is characterized by high summer heat and very cold winters. Precipitation is less than in the mountains and for the Mediterranean climate—and over a wide range, with from 600–1,200 mm annual rainfall. This is also expressed by the occurrence of frosts. In the summer, shorter or longer periods of dry weather occur (Gajić-Čapka 2004).

In the Mountainous region, at higher altitudes of the Dinaric mountains, Gorski Kotar, and Lika, there is a typical mountain climate. Its main feature is the big difference in air temperature between day and night, causing strong physical weathering of rocks. Conversely, the abundance of precipitation leads to rain- and snow-related erosion in the karst causing cracks, karst holes (vrtača), topographic depressions or karst fields. A perhumid and humid climate prevails.

In the Adriatic region a slightly cooler climate dominates, i.e., the northern variant of the Mediterranean climate, with significant differences between the northern and southern subregion. The main climatic features are hot and dry summers and mild rainy and windy winters. The greatest rainfall is in late autumn. In the summer, heat and dry months lead to high evapotranspiration. The dominant winds are bura and jugo. Both cause strong wind erosion of soil and/or weathered rock fragments or humus layer, if not protected. We remember for example the year 2012, when in the Vinodol valley a wind bura swept the soil and planted vines, and the soil was removed along with its crop of barley in Čepić field, near Labin.

One of the most important climatic elements that affect weathering of rock is temperature. The mean annual air temperature in the Croatian territory ranges from 3 °C in the highest mountains to 17 °C in the coastal area. The maximum mean monthly temperatures occur in July, less frequently in August, and the minimum usually in January. The spatial distribution of mean annual air temperature in Croatia is shown in Fig. 4.2.

Water participates in practically all chemical processes of soil formation, because they take place in the aqueous solutions of regolith or soil. Firstly, after precipitation the detritus and regolith receives the water, and after that the received water supplies the plant and soil biosphere and some is passed through the soil mass (macropores), at which point some components are leached out. These are first the easily soluble salts (chlorides, nitrates), then the moderately soluble salts, and after that comes acidification followed by soil eluviation and illuviation of clay minerals, migration, and then the destruction and leaching of components of clay minerals. The average annual rainfall varies from 450 to over 3,500 mm, and its spatial distribution is shown in Fig. 4.3.

As can be seen, minimal precipitation occurs in the eastern Pannonian subregion, where the results of our research have shown frequent droughts in recent years and the process of desertification. In the mountain subregion (G-1) and the Adriatic agricultural region minimum precipitation falls in the warm summer month of July, while the maximum rainfall usually occurs in November. The Pannonian region is characterized by minimal precipitation in the cold season (January or February), while the maximum precipitation falls in the warm months of the year (mostly in June). Since in that period the crop does not cover the soil surface or is just sprouting, rainfall of high intensity causes strong water erosion, especially on sloping fields. For example, this is the case for maize, the prevalent crop on Croatian fields; erosion removes the surface layer of topsoil mass, which is fertilized before sowing and treated with herbicides, thus causing damage to aquatic ecosystems, eutrophication, and biocide effects on the biological complex of water. This means redistribution of fertilizers and herbicides and higher concentrations in the soil mass at the foot of hills, which gradually becomes the source of water eutrophication and/or pollution.

4.2.1 Pannonian Region

The Pannonian region, the most important agricultural region, shows differences in climatic features between its subregions. Using Rfm (Rainfall factor monthly) according to Gračanin (1950), we analyzed hydrothermic conditions and water balance after Thornthwaite—Table 4.1.

All climatic indicators manifest certain changes on going from the east to the west. The eastern part is characterized by a semiarid continental climate, which means dry and torrid summers, and harsh winters, in contrast to the western

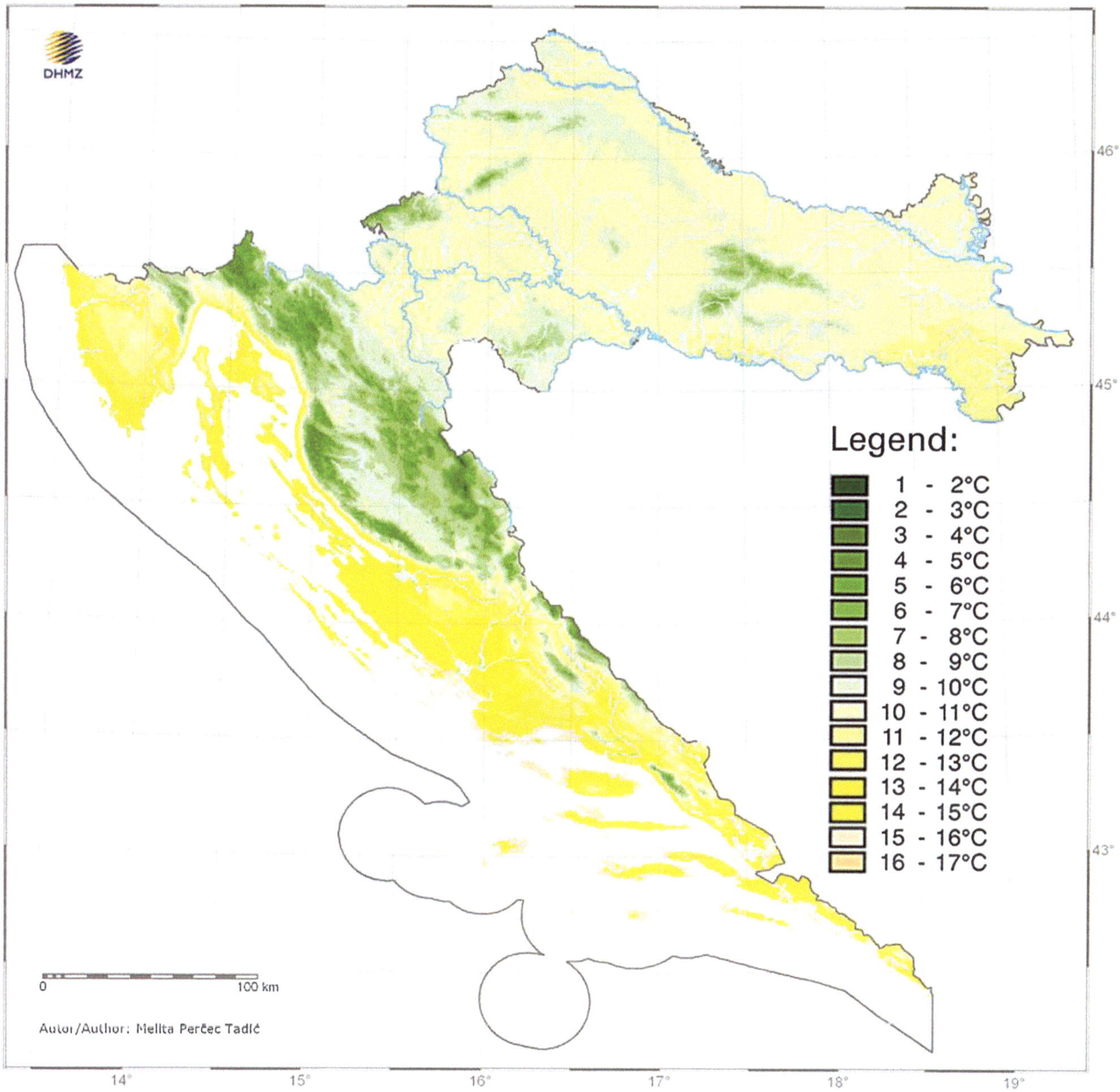

Fig. 4.2 Mean annual air temperature in °C (Zaninović et al. 2008)

part with a humid climate with considerably higher precipitation and smaller temperature amplitudes (Gajić-Čapka 2004).

As can be seen, the average monthly precipitation gradually increases going from east to west; in Ilok, it amounts to 640 mm, in Slavonski Brod 781, Zagreb 854, and in Varaždin 840 mm. The mean annual temperature decreases in the same direction, from 11.1 in Ilok to 9.9 °C in Varaždin. Precipitation distribution is relatively favorable, most falling in the growing period with a maximum in June. Some of these rains can be torrential, especially in the P-1 subregion, which causes very intensive erosion on short slopes of loess dunes and forms bogs in depressions between dunes. In Ilok, however, all the summer months from July to September are arid, while April, May, June, and October are semiarid. Conversely, in Varaždin July is the only arid month, and there are no arid months in Slavonski Brod and Zagreb. The data on water balance are also interesting. Ilok has an water deficiency of 108 mm in the summer but a surplus of 115 mm in the winter. Slavonski Brod has 50 mm of deficit Zagreb is without deficit but in Varaždin, it is a symbolic 13 mm. Very

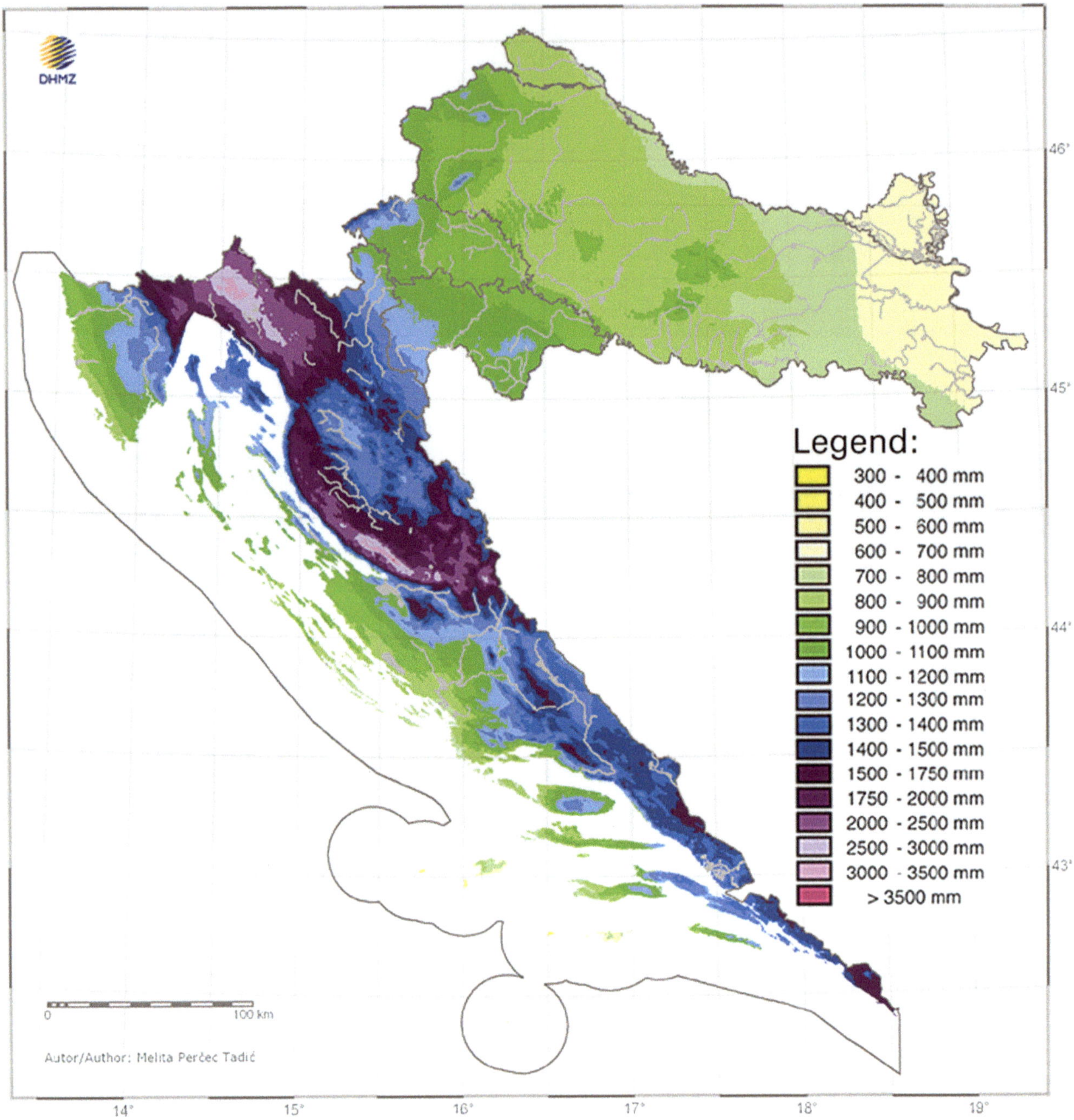

Fig. 4.3 Map of the mean annual precipitation for the period 1961–1990 (Zaninović et al. 2008)

influential for soil genesis is water surplus, which is 150 mm in Slavonski Brod, in Zagreb 185 but in Varaždin 200 mm. This water is the main factor of leaching of different soil components—in P-1 of moderately soluble salts ($CaCO_3$, $MgCO_3$) and decarbonatization, in P-2 acidification and partly eluviation of clay, in P-3 eluvial–illuvial migration of clay and water stagnation, and in P-4 erosion of soil removing the surface horizon and water stagnation on compacted impermeable soil horizons.

4.2.2 The Mountainous Region

The Mountainous region comprises a relatively heterogeneous region, starting with the so-called shallow or covered karst of the Karlovac hinterland, continuing towards the mountain massif of Mala and Velika Kapela, Gorski Kotar, Velebit, and the Dinara mountain massif. In this direction the mean annual precipitation increases, and it is at a maximum on the tops of the mountains. In Table 4.2 we

Table 4.1 Indicators of climate conditions in the Pannonian agricultural region

Climatic data	Month												Annual
	I	II	III	IV	V	VI	VII	VIII	IX	X	XI	XII	
Ilok (based on data for 1961–1990) P-1 subregion													
Precipitation (mm)	42.6	43.2	46.5	51.4	59.9	82.6	65.2	58.2	43.2	41.1	53.3	54.4	641.7
Mean temperature (°C)	−0.5	2.2	6.5	11.5	16.3	19.2	20.9	20.5	17.0	11.8	6.3	1.7	11.1
Monthly rain factor (Rfm)	–	19.6	7.1	4.5	3.7	4.3	3.1	2.8	2.5	3.5	8.5	32	58
Humidity of cl imate[a]	–	ph	h	sa	sa	sa	a	a	a	sa	h	ph	SA[b]
Potential evapotr. (mm)	0.0	4.9	24.0	52.4	92.8	1186	131.1	1183	80.0	46.3	189	4.0	691.3
Real evapotranspir. (mm)	0.0	4.9	24.0	52.4	92.8	118.6	109.1	64.5	47.6	46.3	18.9	4.0	583.0
Water deficit (mm)	0.0	0.0	0.0	0.0	0.0	0.0	22.1	53.8	32.5	0.0	0.0	0.0	108.3
Water surplus (mm)	46.3	43.2	21.7	4.1	0.0	0.0	0.0	0.0	0.0	0.0	0.0	0.0	115.3
Slavonski Brod (based on data for 1965–1984) P-2 subregion													
Precipitation (mm)	49.0	49.0	48.0	61.0	75.0	92.0	89.0	74.0	58.0	55.0	65.0	67.0	781
Mean temperature (°C)	−1.0	1.8	5.9	10.8	15.7	18.8	20.5	19.8	16.3	11.2	5.6	1.1	10.5
Monthly rain factor (Rfm)	–	26.5	8.1	5.6	4.8	4.9	4.4	3.7	3.6	4.9	11.7	62.2	74
Humidity of climate[a]	–	ph	h	sh	sa	sa	sa	sa	sa	sa	h	ph	SH
Potential evapotr. (mm)	0.0	4.4	225	52.1	92.8	117.2	131.3	116.2	78.7	45.1	16.3	2.1	678.6
Real evapotranspir. (mm)	0.0	4.4	22.5	52.1	92.8	117.2	131.3	88.8	58.1	45.1	16.3	2.1	630.7
Water deficit (mm)	0.0	0.0	0.0	0.0	0.0	0.0	0.0	27.3	20.6	0.0	0.0	0.0	47.9
Water surplus (mm)	49.1	44.4	25.1	8.4	0.0	0.0	0.0	0.0	0.0	0.0	0.0	23.5	150.6
Zagreb–Maksimir (based on data for 1965–1984) P-3 subregion													
Precipitation (mm)	48.7	41.4	55.4	58.3	76.2	98.2	89.5	94.9	77.4	87.5	77.7	63.8	854.0
Mean temperature (°C)	−0.2	2.2	6.3	10.2	15.2	18.6	20.0	19.0	15.6	10.4	5.2	1.0	10.3
Monthly rain factor (Rfm)	–	18.8	8.8	5.7	5.0	5.3	4.5	4.9	4.9	7.0	14.9	58.0	83
Humidity of climate[a]	–	ph	h	sh	sa/sh	sh	sa	sa	sa	h	ph	ph	H
Potential evapotr. (mm)	0.0	6.0	25.0	50.0	91.0	117.0	128.0	112.0	76.0	42.0	16.0	2.0	666.0
Real evapotranspir. (mm)	0.0	6.0	25.0	50.0	91.0	117.0	128.0	112.0	76.0	42.0	16.0	2.0	666.0
Water deficit (mm)	0.0	0.0	0.0	0.0	0.0	0.0	0.0	0.0	0.0	0.0	0.0	0.0	0.0
Water surplus (mm)	49.0	35.0	30.0	12.0	0.0	0.0	0.0	0.0	0.0	0.0	0.0	60.0	185.0
Varaždin (based on data for 1965–1984) P-4 subregion													
Precipitation (mm)	48.2	43.7	51.1	64.7	81.6	100.3	96.1	48.7	79.6	65.8	91.3	69.0	840.1
Mean temperature (°C)	0.8	1.7	5.8	9.9	14.8	17.4	19.7	19.1	15.3	10.1	4.8	0.6	9.9
Monthly rain factor (Rfm)	60.2	25.7	8.8	6.5	5.5	5.8	4.9	2.5	5.2	6.5	19.0	11.5	84.8
Humidity of climate[a]	ph	h	sh	sh	sh	sa	a	sh	sh	ph	ph	ph	H
Potential evapotr. (mm)	0.0	4.7	24.3	49.6	89.7	109.6	127.3	113.5	75.1	42.2	15.2	1.3	652.5
Real evapotranspir. (mm)	0.0	4.7	24.3	49.6	89.7	109.6	127.3	100.2	75.1	42.2	15.2	1.3	639.2
Water deficit (mm)	0.0	0.0	0.0	0.0	0.0	0.0	0.0	13.4	0.0	0.0	0.0	0.0	13.4
Water surplus (mm)	48.2	39.0	26.8	15.1	0.0	0.0	0.0	0.0	0.0	0.0	4.2	67.7	200.9

Criteria for characterization of humidity:

[a] For monthly humidity: Monthly rain factor—Rfm (Mean monthly precipitation in mm: mean monthly temp. in °C). After Gračanin (1950): *a* arid (1.6–3.3), *sa* semiarid (3.3–5.0), *sh* semihumid (5.0–6.6), *h* humid (6.6–13.3), *ph* perhumid—over 13.3

[b] For annual humidity: using the Rain factor of Lang—RF (Annual precipitation in mm: Annual temperature °C) *A* arid 0–40, *SA* semi arid 40–60, *SH* semihumid 60–80, *H* humid 80–160, *PH* perhumid over 160

present the main climatic indicators affecting soil genesis conditions for this region, based on the data of Karlovac for G-1 and Parg as the highest positioned station at 863 m a.s.l. (45°36′N, 14°38′E) near Čabar in Gorski kotar for G-2.

The climate is typically mountainous, with very high precipitation, a significant part of which falls in the form of snow, so that Karlovac is humid but Parg perhumid. A characteristic feature is the comparatively short growing

Table 4.2 Indicators of climate conditions in the Mountainous agricultural region

Climatic data	Month												Annual
	I	II	III	IV	V	VI	VII	VIII	IX	X	XI	XII	
Karlovac (based on data for 1965–1984) G-1 subregion													
Precipitation (mm)	76.3	67.4	86.9	90.9	119.5	97.1	97.5	105.8	97.1	101.1	121.7	96.6	1157.8
Mean temperature (°C)	−0.4	2.1	6.4	10.9	15.7	19.3	21.1	20.0	16.3	11.1	6.0	1.2	10.8
Monthly rain factor (Rfm)	–	31.6	13.7	8.3	7.6	5.0	4.6	5.3	6.0	9.1	20.3	82.1	107
Humidity of climate[a]	–	ph	ph	h	h	sh	sa	sh	sh	h	ph	ph	H
Potential evapotr. (mm)	0.0	5.1	24.1	51.9	92.2	120.6	135.0	116.9	77.5	44.1	17.4	2.2	687.0
Real evapotranspir. (mm)	0.0	5.1	24.1	51.9	92.2	120.6	135.0	116.9	77.5	44.1	17.4	2.2	0.0
Water deficit (mm)	0.0	0.0	0.0	0.0	0.0	0.0	0.0	0.0	0.0	0.0	0.0	0.0	0.0
Water surplus (mm)	104.3	94.3	62.8	38.9	27.3	0.0	0.0	0.0	0.0	4.5	104.3	94.3	470.9
Parg (based on data for 1965–1984) G-2 subregion													
Precipitation (mm)	133.1	131.4	129.2	146.6	142.6	155.9	131.2	135.9	178.8	215.3	235.4	175.0	1903.8
Mean temperature (°C)	−2.0	−1.2	1.8	5.7	10.6	14.1	16.2	15.3	12.3	8.0	3.3	−0.6	7.0
Monthly rain factor (Rfm)	–	–	73.1	25.7	13.5	11.0	8.1	8.9	14.5	27.0	72.1	–	273
Humidity of climate[a]	–	–	ph	ph	ph	h	h	h	ph	ph	ph	ph	PH
Potential evapotr. (mm)	0.0	0.2	17.0	43.2	80.3	104.7	118.2	104.0	69.8	41.2	16.8	0.0	595.4
Real evapotranspir. (mm)	0.0	0.2	17.0	43.2	80.3	104.7	118.2	104.0	69.8	41.2	16.8	0.0	595.4
Water deficit (mm)	0.0	0.0	0.0	0.0	0.0	0.0	0.0	0.0	0.0	0.0	0.0	0.0	0.0
Water surplus (mm)	108.3	108.6	87.8	67.1	27.0	0.0	0.0	0.0	0.0	120.1	149.4	162.0	830.3

See Table 4.1 footnote

period due to late spring and early autumn frosts. With forest and/or mountainous pastures as the dominant natural vegetation, this geographical area represents a typical forest region.

Average precipitation in Karlovac is 1,157, but in Parg 1,904 mm; in Karlovac July is semiarid, and the other summer months are semihumid; autumn and winter are perhumid. But in Parg all months are perhumid except for the period June–August, which is humid. Winter precipitation is in the form of snow. Water surplus is very high—in Karlovac 471 mm and in Parg 830 mm. Processes of leaching are dominant, acidification caused by fulvic acids from humus is very high, causing destruction of clay minerals and leaching of their components—podzolization. In the soil profile siliceous sand of gray color is visible in the eluvial horizon in contrast to the rusty illuvial horizon. Water from the entire area belongs to the Kupa river catchment area, and while the landscape is intersected by beautiful rivers, the real "karst beauties" of irresistible charm include the Dobra, Mrežnica, and Korana rivers, to say nothing of the exceptional world heritage site of Plitvice Lakes, protected by UNESCO (Vrbek and Pernar 2001).

4.2.3 Adriatic Region

The Adriatic agricultural region has all the characteristics of a Mediterranean climate, and its indicators are shown in Table 4.3.

The presented climatic conditions of the Adriatic agricultural region show a difference going from north-west to south-east; the mean annual temperature shows a steady increase and so does precipitation. The climate is warm, with abundant sunshine, temperatures only rarely drop below 0 °C, and the summer months (including September in Dubrovnik) are torrid. Three winter months are moderately cold in Pula (J-1); only two are cold in Zadar (J-2), while Dubrovnik (J-3) has a moderately warm winter period. Snow is very rare. Hence, the winter months are perhumid or humid, and the summer months arid, except for Dubrovnik, where July is perarid. Under such conditions, two opposites prevail regarding the water balance—a marked surplus, causing erosion in winter, and a large deficit during the summer months, which calls for irrigation.

The water balance shows a relatively high water deficit in the summer period of 120 mm in Pula, 150 mm in Zadar,

Table 4.3 Indicators of climate conditions in the Adriatic agricultural region (based on data for the period 1965–1984)

Climatic data	Month												Annual
	I	II	III	IV	V	VI	VII	VIII	IX	X	XI	XII	
Pula—North Adriatic subregion - J-1													
Precipitation (mm)	81.7	72.1	57.9	68.2	65.0	54.0	39.8	98.2	97.4	70.0	109.9	87.4	901.6
Mean temperature (°C)	5.5	6.1	8.4	11.7	16.3	20.4	23.1	22.2	18.7	14.4	9.9	6.7	13.6
Monthly rain factor (Rfm)	14.8	11.8	6.9	5.8	3.9	2.6	1.7	4.4	5.2	4.9	11.1	13.0	66.3
Humidity of climate[a]	ph	h	h	sh	sa	a	a	sa	sh	sa	h	h	SH
Potential evapotr. (mm)	11.7	13.7	26.9	47.7	85.9	120.1	144.4	126.0	85.6	53.9	26.6	14.5	757.1
Real evapotranspir. (mm)	11.7	13.7	26.9	47.7	85.9	120.1	52.8	98.2	85.6	53.9	26.6	14.5	637.7
Water deficit (mm)	0.0	0.0	0.0	0.0	0.0	91.7	27.8	0.0	0.0	0.0	0.0	0.0	119.4
Water surplus (mm)	70.0	58.4	31.0	20.5	0.0	0.0	0.0	0.0	0.0	0.0	11.2	72.9	263.9
Zadar—Central Adriatic subregion - J-1													
Precipitation (mm)	80.3	74.7	74.0	62.1	64.8	57.1	35.1	75.3	102.5	103.3	112.5	100.5	942.0
Mean temperature (°C)	6.9	7.5	9.5	12.6	17.1	20.9	23.4	22.9	19.7	15.6	11.5	8.1	14.6
Monthly rain factor (Rfm)	11.6	9.9	7.8	4.9	3.8	2.7	1.5	3.2	5.2	6.6	9.8	12.4	64.5
Humidity of climate[a]	h	h	h	sa	sa	a	a	a	sh	sh/h	h	h	SH
Potential evapotr. (mm)	14.2	16.3	29.1	49.4	882	121.4	145.2	129.7	89.5	57.4	30.5	17.0	788.1
Real evapotranspir. (mm)	14.2	16.3	29.1	49.4	882	121.4	47.3	75.3	89.5	57.4	30.5	17.0	635.8
Water deficit (mm)	0.0	0.0	0.0	0.0	0.0	0.0	97.9	54.4	0.0	0.0	0.0	0.0	152.3
Water surplus (mm)	65.8	58.4	44.9	12.7	0.0	0.0	0.0	0.0	0.0	0.0	40.9	83.5	306.2

(continued)

Table 4.3 (continued)

Climatic data	Month												Annual
	I	II	III	IV	V	VI	VII	VIII	IX	X	XI	XII	
Dubrovnik—South Adriatic subregion -J-3													
Precipitation (mm)	138.0	121.0	99.0	89.0	67.0	54.0	32.0	93.0	103.0	111.0	138.0	134.0	1179.0
Mean temperature (°C)	8.8	9.3	11.2	13.8	18.1	21.8	24.1	24.1	21.4	17.5	13.6	10.3	16.2
Monthly rain factor (Rfm)	15.7	13.0	8.8	6.4	3.7	2.5	1.3	3.9	4.8	6.3	10.1	13.0	72.7
Humidity of climate[a]	ph	h	h	sh	sa	a		sa	sa	sh	h	h	SH
Potential evapotr. (mm)	17.5	19.4	32.9	50.7	89.7	123.4	147.1	136.7	98.0	63.9	36.2	21.9	837.2
Real evapotranspir. (mm)	17.5	19.4	32.9	50.7	89.7	123.4	39.9	93.0	98.0	63.9	36.2	21.9	686.4
Water deficit (mm)	0.0	0.0	0.0	0.0	0.0	0.0	107.1	43.7	0.0	0.0	0.0	0.0	150.8
Water surplus (mm)	120.5	101.6	66.1	38.3	0.0	0.0	0.0	0.0	0.0	0.0	53.9	112.1	492.6

See Table 4.1 footnote

but 492 mm in Dubrovnik. Conversely, in the winter period there is a water surplus of 250–500 mm.

Water plays an important role in the processes of formation and evolution of soil by its involvement in the physical weathering of parent rocks. It thus forms coarse regolith, moving it down the slope by erosion and causing its fragmentation. Then follows chemical weathering, the formation of clay minerals, supply of plants with nutrients from the regolith/soil solution, and leaching of easily soluble salts. During soil evolution in the first phase we have slight, but, in the next phase, intensive acidification, followed by eluvial–illuvial migration, and in the last phase destruction of clay minerals by the process of podzolization. Water from rainfall participates in the translocation of regolith/soil through the process of water erosion. Stagnation of water on some horizons of the soil profile or in the whole profile of the soil causes hydromorphic processes, which means chemical processes of the reduction of some soil substances and transformation of organic matter in the reducing conditions by anaerobic microflora. Genetic horizons of soils saturated with water limit root penetration and reduce the physiologically active soil depth.

The origin, duration, and intensity of the excess moisture determine the needs and opportunities for hydrotechnical measures of water drainage.

The wetting regime, i.e., the origin and effect of water on the soil serves as a basis for soil classification to Soil order as the highest category in a hierarchical system of soil classification in Croatia. Subsequent Soil orders are defined in the classification of soils of Croatia as follows (Husnjak 2012):

Soil order of terrestrial (automorphic) soils. Here we have soils wetted by precipitation only. Surplus water (over field capacity of soil) free and without prolonged retention trickles through the soil profile. Excessive wetting by superfluous water does not appear within 1.0 m of soil depth in this soil.

Soil order of semiterrestrial soils. These are characterized by occasionally excessive wetting by superfluous water (within 1.0 m of soil depth), whose origin is exclusively rainwater, and which is "stagnant" in and/or on a compacted horizon of moderately poor to poor water permeability.

Soil order of hydromorphic soils. These are characterized by excessive wetting by groundwater within 1.0 m of soil depth, and additional water, whose origin may be flood or rainfall, which has stagnated for a long time on/in a horizon of low to very low permeability.

Soil order of halomorphic soils. These are also characterized by excessive wetting by superfluous water (within 1 m soil depth), which is salinized and/or is alkaline.

Soil order of subaquatic soils. These are characterized by origin and development that occurs under shallow water cover or under permanent underwater conditions.

4.3 Biological Factors of Soil Genesis

Soil is formed on the detritus resulting from weathering of debris of consolidated or nonconsolidated parent material. But the real beginning of soil genesis is considered to be the moment when the first higher organism occurs within the soil, or when the regolith as the initial soil begins biological cycling of matter and energy and becomes part of the biosphere—the phytosphere. From that moment on a relationship of mutual action between soil and organisms is established: the action of soil on organisms (pedobiosphere), and of organisms on soil (pedosphere). Organisms affect the soil through exchange of materials and energy; they are adapted to the soil, influence it, and change its physical, chemical, and biological properties and in this way open the door to the next phase of the soil evolutionary series.

The first higher organisms that inhabit the regolith are lichens, followed by more modest-demand plants that, like lichens, leave a significant amount of soil organic matter, humus, as a colloidal complex. Humus holds more water, forms a cation exchange system in water solution, binds nutrients, and after its decomposition by mineralization releases nutrients for new growing plants. Of special importance are solely automorphic organisms—the vegetation, or phytosphere. Such organisms that are found in the pedosphere are an inseparable part of the soil, and they are joined by other organisms and biocenoses living in or on the soil surface or partly in the soil, and together, in interdependence with other habitat factors, they direct the processes of soil genesis and the evolution of soil. Therefore, natural biocenosis indicates soil and habitat state, including climate and relief conditions. The surface of the Earth was initially covered by natural plant communities only—phytocenoseis—whose structure depends directly on soil characteristics. Expansion of the agrosphere is always done at the expense of soils under natural phytocenosis. Croatia no longer has any of these soils, but we have just the opposite: there are areas of marginal soil cultivated under the different socio-economic conditions of today, which are no longer usable for agriculture and which should be restored to the original—mostly forest vegetation.

Vegetation affects the formation of soil by synthesis of organic matter and accumulation of humus, which forms the A horizon of the soil profile. In addition, respiration of the rhizosphere releases CO_2, which with water forms carbonic acid, an important agent of chemical weathering of soil

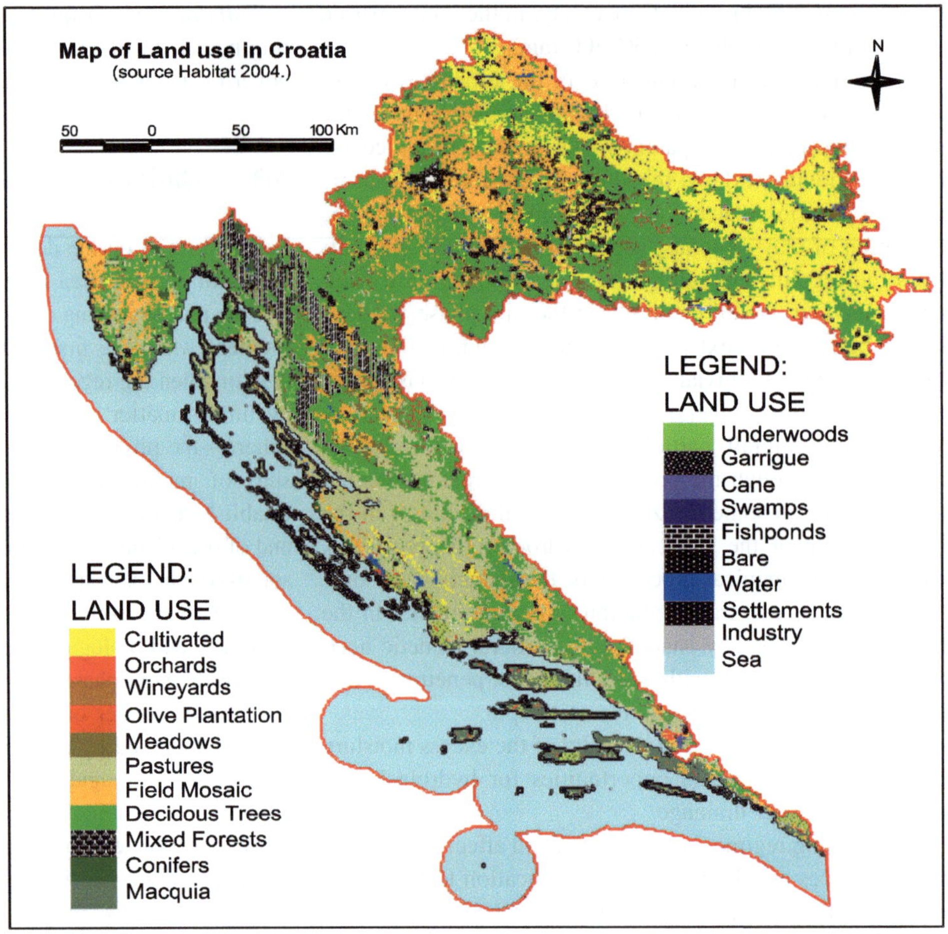

Fig. 4.4 Land use map (Map of habitat, 2004)

minerals. Grass vegetation, through its roots, has a positive effect on the formation of a stable crumb structure, and in this way adds favorable water–air relation and permeability to water. Vegetation roots take up plant nutrients from deeper layers of the soil, accumulate it in the surface layer, and this protects against permanent loss from the soil. Vegetation also affects the microclimate and pedoclimate.

According to the heterogeneity already described of parent material, climate, and relief with its various forms and expositions, the main feature of the biosphere of Croatia is its very high diversity, which is illustrated by the data on about 2,250 plant varieties on Velebit mountain. The structure and surface modes of use of agricultural and forest land are presented in Fig. 4.4.

The areas of different forms of land use are given in Table 4.4.

There are significant differences in impact on the formation and evolution of natural soils between spontaneous phytocenosis and agrobiocenosis, low-growing and high-growing plants, and annual and perennial plant species.

4.3.1 Steppe Phytocenosis

Steppe plant communities are characterized by high participation, or even prevalence of grass species (Gramineae) in the botanical composition. Unlike the forest, the grass vegetation forms a very dense system of fine roots and so we have a relatively large depth of soil enriched with organic matter. Because of climatic conditions (arid climate) there is no washing out, with vegetation residues forming a high-quality, base-saturated humus. Steppe vegetation returns to the soil plant nutrients, biogenic elements, by returning organic residues accumulated during growth. Few survive in Croatia, for example on the hill Branjsko brdo in Baranja and on the edges of forests.

4.3.2 Meadow Phytocenosis

Meadow plant communities represent a grass phytocenosis of humid climates. In contrast to the steppe, meadow

Table 4.4 Structure of agricultural and woodland

Land use	Area	
	ha	(%)[b]
Agricultural land		
Tilled area	878,319	18.0
Pastures	786,236	16.1
Field mosaic[a]	689,544	14.1
Meadows	239,198	4.9
Vineyards	19,267	0.4
Orchards	7,047	0.1
Olive plantations	5,751	0.1
Total agricultural land area	2,625,362	53.7
Woodland		
Deciduous trees	1,917,262	39.2
Maquis	156,869	3.2
Garrigue	90,570	1.9
Conifers	56,754	1.2
Mixed forests	23,260	0.5
Underwood	21,957	0.4
Total woodland area	2,266,672	46.3

[a] Field mosaic—cultivated land including agricultural plots used as tilled area and gardens, vineyards, orchards, olive plantations, but because of small plots impossible to present as separate units
[b] Percent in relation to sum of agricultural and forest land

vegetation forms a small amount of humus as it is already being decomposed and the humus is of inferior quality, because of greater leaching of bases and an acid soil chemistry. Hydrophilic plant communities grow under conditions of excessive moisture. This vegetation provides a large amount of organic residues, which are slowly and incompletely degraded due to the anaerobic conditions. Thus, they form gley and peat soils with a specific type of hydromorphic humus.

Cultured plants are mostly annual species. Unlike natural phytocenosis, agrobiocenosis is poor in species, because man removes all other species except those that are his breeding goal. Equilibrium between plant species in the spontaneous phytocenosis is established because of very diverse allelopathic relations. In the agrobiocenosis there is no spontaneous equilibrium between varieties; the relations are established, maintained and directed by man, without which the relation would not be upheld. Our calculations (Bašić and Herceg 2009) show that the Croatian and neighboring agrosphere regularly breed over 65 species from 13 botanical families, of which 32 species are the most common Fabaceae, while the Poaceae, represented by 14 species, are prevalent according to area occupied, followed by Chenopodiacea and Brassicaceae. Another difference compared with natural vegetation is that crops are harvested with irreversible loss of nutrients, which form part of the composition of the biological mass of the yield. The crop protects the soil from evaporation, and some species, for example legumes, enrich the soil with nitrogen, and a perennial clover–grass mixture, because of the large mass of its roots in the surface layer, has a positive influence on the formation of a granular soil structure.

4.3.3 Forest Phytocenosis

Forest phytocenosis is perennial plant cenosis in semihumid and humid climates. Only a small part of the aerial organs of forest trees die each year, creating a surface cover in the forest—litter (O horizon). Therefore, the amount of humus in soils under forest vegetation rapidly decreases with depth. The root mass of trees is 20–30 % of the total plant mass. Residues of forest vegetation contain a large amount of biogenic elements—plant nutrients that are released in a microbiological process of transformation of organic matter—humification and mineralization of humus. A higher content of nutrients is found in residues of vegetation of deciduous forests than in conifers, the chemical composition of which has a high content of lignin and tannin, which is reflected in the type and quality of humus. These are the remains of acid-forming humus, rich in fulvic acids. According to Vukelić and Rauš (1998) residues of beech

(*Fagus silvatica*) have weakly acid humus, but the ash (*Fraxinus excelsior*), common maple (*Acer campestre*), elm (*Ulmus minor*), and linden (*Tillia* sp.) have almost neutral humus of high quality. In forest soils, with the same quantity of precipitation, leaching of soil is much stronger than in soils under grass vegetation. Intense leaching is a consequence of increased descending flows of water under the influence of the roots of the forest vegetation that strongly dry deeper soil layers, thus accelerating the descending movement of water.

Forest influences the thermal regime in such a way that the soil under forest vegetation is an average of 2–3 °C cooler than bare soil. These differences are far greater in the summer. Annual and daily (day–night) amplitudes of temperature in soils under forest vegetation are far less. Forest vegetation significantly changes the water regime of the soil. Part of the water from rainfall is retained in the aerial parts, the crown of trees, and re-evaporated into the atmosphere, but part of the water that falls to the soil surface completely infiltrates the litter. Evaporation from the bare surface of soil is less than half that of the forest surface due to lower soil and air temperatures, poor air flow, and higher relative humidity in forests. Conversely, there are large losses of water from the soil via roots for transpiration. As the roots penetrate deeper and more intensely into the deeper layers, these become dry and this intensifies descending flows of water and increases soil leaching, acidification and eluvial–illuvial migration of clay mineral. An example: near Vukovar within the forest, a loess soil type is in the cambisol eutric phase of evolution, but outside the forest there is a leached chernozem with the first signs of a cambic process (Bogunovic et al 1984).

Vukelic and Rauš (1998) found in Croatia about 60 major forest communities classified into two major vegetation regions of the world, whose natural boundary lies just in the Croatian territory—on the slopes of the coastal Dinarides. The Mediterranean region belongs to the lower forests of downy oak (*Quercus pubescens*), a Siberian–North-American region at higher altitude to the beech (*Fagus sylvatica*) forest.

4.3.3.1 Eurosiberian–North-American Vegetation Region

The Eurosiberian–North-American vegetation region, a European subregion, varies with lowlands, hills, hilly, mountainous, and subalpine vegetation belts, and each of them has several vegetation zones. The Lowland vegetation zone is the backbone of forest vegetation in the Pannonian region. In the semiarid area of the eastern Pannonian subregion (P-1) Croatian scientists distinguish a real Pannonian vegetation zone, a forest transitory to steppe, and a steppe proper. In the Danube region and smaller tributaries of the river Drava swamp forests are found in which occur white willow (*Salix alba*), gray alder (*Populus nigra* and *alba*), and other species that are representative of regularly flooded habitats.

In the floodplain Vukelić et al. (2008) found 15 forest communities within two types of flooded forest:

Coastal floodplain forests of willows and poplars

This community occurs near Vraždin, in periodically flooded areas on sandy and gravelly soils in the old bed of the river Drava, in the basins of the Danube in Baranja, Drava and Mura, and on islands on the Danube. Dominant varieties are *Salix alba, fragilis, purpurea, cinerea, triandra, europaeus, Populus alba, Populus nigra, Alnus incana, Alnus glutinosa*, accompanied by *Rubus hirtus.*

Floodplain forests of pedunculate oak, black alder, and narrow-leaved ash

This community occurs along the course of the Drava River, the Sava with Bosut as confluents, in Lonjsko polje, and in the famous forest Spačva, with small fragments in the lowland forests of Đurđevac and in the Pokupsko. It indicates short-lasting periodic flooding of sandy fluvial soils and hypogley. Indicator varieties are: *Quercus robur* as the leading and most valuable forest variety, followed by *Fraxinus angustifolia, Ulmus* and *Ulmus carpinfolia, Acer tataricum, Crataegus nigra, Cornus sanguinea, Prunus padus*, and sporadic *Fraxinus angustifolia* and *Quercus robur.* We have the well-known relict forest "Motovun forest" situated in the valley of the river Mirna in Istria with *Quercus robur*, and *Carpinus betulus*, while *Ulmus carpinifolia* is less common.

At somewhat higher, drier positions in the valley occur forests of common oak (*Quercus pedunculata*) and hornbeam (*Carpinus betulus*) (Photo 4.21).

In addition there is alder (*Populus nigra*), willow (*Salix* sp.), and poplar (*Populus nigra*), whose survival is more or less related to the duration of the excess moisture regime of surface and groundwater. For growth and for these species and forest communities to thrive, river valleys are the most appropriate where in the meso-relief exchange ponds, depressions, and moist broad terraces occur. Soils are the most common mineral-marsh, gley, but on the slope of the hills pseudogley occurs. The excess moisture, the wetting regime (related to the origin and intensity of sufficient wetting), is a vital factor in the emergence, development, and structure (participation of tree varieties) of a lowland forest vegetation zone. Floodwater affects the formation of poplar (*Populus nigra*), and willow (*Salix* sp.) forests, underground water affects the forests of common oak (*Quercus pedunculata*); and forests dominated by red alder (*Rhamnus frangula*), and ash (*Fraxinus angustifolia*) were formed under the occasional influence of both types of water.

Photo 4.21 Forest community of common oak and hornbeam in Pokuplje basin (Photo 4.Vrbek)

Forests of deciduous oaks outside the reach of floods

This forest is widespread in the Pannonian part of Croatia: Posavina, Podravina, Pokuplje and central Croatia, Slavonia, western Srijem, and Baranja. Dominant varieties are *Quercus cerris* and *Quercus petraea, Carpinus betulus, Acer campestre, Prunus avium*, and *Tilia cordata*, widespread on the western slopes of Fruška gora up to an altitude of 250 m. *Tilia tomentosa* has secondary importance. On the eastern hills of Kalnik, and on Bilogora, Papuk, Požeška gora, and Dilj there are forests of sessile oak and common hornbeam *Carpinus betulus.*

Forests of sessile oak and sweet chestnut are found on Zrinjska Gora, Medvednica, then on Papuk, Psunj, Požeška Gora, and Šamarica on shallow acid soils on metamorphic rocks—schists and sandstones.

The hilly belt of vegetation in the lowland continues and extends to a height of 150 to 500 m a.s.l., where pedoclimatic (soil and climate) conditions provide favorable conditions for forest vegetation. Therefore, forest communities are lush and species rich. Since they developed, because of the very favorable conditions for human life, in populated areas, they were in the past used for acorn fodder in extensive farming with the rearing of free-running pigs, or as pasture for cattle. These forests are generally strongly under anthropogenic influence. Given the relatively favorable characteristics of soils, man cleared the forest and cultivated the land. Overgrown hills and high mountains where they come around the lower part of such mountains as a ring, as is the case with the Slavonian mountains (Psunj, Papuk, Dilj, and Krndija), Moslavaka gora, Medvednica, and Ivanšćica. The main tree species in this zone is the sessile oak (*Quercus sessiliflora*), but, depending on the characteristics of the soils, communities differ so that these valuable forest species occur in acidophilic, neutrophilic–mesophilic and thermophilic–basophilic communities, on different geological substrates and on different soil types. Other important species are hornbeam (*Carpinus betulus*), beech (*Fagus sylvatica*), chestnut (*Castanea sativa*), pubescent (*Betula pubescens*), Turkey oak (*Quercus cerris*), cherry (*Prunus avium*), and other species.

4.3.3.2 The Mediterranean Vegetation Region

The *Mediterranean littoral zone* covers most of the islands, central and southern Dalmatia, and the narrow coastal area. In this zone grow pure and mixed forests of holm oak (*Quercus ilex*) with many degraded areas, and forests of Aleppo pine (*Pinus halepensis*). In the coastal forests the dominant trees are pubescent oak (*Quercus pubescens*) with white hornbeam (*Carpinus orientalis*).

A Mediterranean-montane vegetation belt is widespread on the highest topographic position of the islands, at a height greater than 400 m a.s.l., while in the continental, coastal region it develops above 300 m in the north Adriatic, and above 600 m a.s.l. in the southern part, the so-called epi-Mediterranean vegetation zone. The most important forest species in this community is hornbeam (*Ostrya carpinifolia*), while in the Mediterranean zone the trees on the islands include pubescentoak (*Quercus

Photo 4.22 Specific form of limestone ("rider") of subalpine vegetation zone in Velebit

pubescens) with pedunculate oak (*Quercus pedunculata*) in the continental part. On the flysch part of Istria a forest has developed with *Quercus pubescens*, *Ostrya carpinfolia*, and *Acer obtusatum*.

Dalmatian black pine is significant in some areas (*Pinus nigra* ssp., *Dalmatica*), and on slopes European silver fir (*Abies alba* ssp. *Biokovensis*). A semi-Mediterranean vegetation of evergreen–deciduous forests covers other parts of Hvar, Brac, Korčula, Mljet, and Pelješac peninsula, at an altitude greater than 400 m. This group belongs to the coastal forest community with black hornbeam (*Ostrya-Fagetum*) and relict forests of linden-strikes (*Tilio-Taxetum*) and a variety of red pine relict communities in the Dinarides and on soils on dolomite. Beech (*Fagus sylvatica*) forests of mountain belts are very rich in plant species and belong to the richest and most abundant beech forests in Europe. In addition to climate and soil conditions, the reasons lies in the historical genetic development of beech and its flora from the postglacial period till today.

4.3.3.3 Altimontane Vegetation Zone

This vegetation zone separates the Dinarides and the Pannonian hills at an altitude of 600 (800) to 1,100 m a.s.l. In this zone today, there are world famous rainforests, i.e., Čorkova uvala near the Plitvice lakes, and Devčića tavani Nadžak bilo in northern Velebit.

4.3.3.4 Subalpine Vegetation Zone

The subalpine vegetation zone includes forest communities at altitudes in the Dinarides of approximately 1,100 m a.s.l. to 1,700 m a.s.l. with a geological-lithological structure dominated by limestone, which for the most part, especially in the spruce (*Picea excelsa*) forests, erupt to the surface, giving this part of the Croatian karst a distinctive appearance, as in Photo 4.22.

The soil types at lower altitudes are kalkocambisol, in the karst limestone-dolomite kalkomelanosol, and rendzina, and lithosol. Anthropogenic influences in this area can be barely noticeable. The reasons are manifold, but mostly lie in its inaccessibility and the harsh climatic conditions unfavorable to human life.

Maquis (macchia)

Maquis are degraded forests of holm oak (*Quercus ilex*) up to 5 m high in which the tree layer is completely lost due to fires or excessive cutting. This shrubland biome of the Mediterranean region in Croatia covers more than 150,000 ha of land in the Adriatic agricultural region. As a rule it consists of densely growing shrubs with xerotherm, drought-resistant plant species such as Christ's thorn (*Paliurus spina-christi*) dominant (Photo 4.23), but also with holm oak (*Quercus ilex*), strawberry (*Rubus ulmifolius*), juniper (*Juniperus oxycedrus*), myrtle (*Myrtus communis*), tree heather (*Erica arborea*), and different climbing plants like wild climber (*Clematis vitalba*). Spanish broom (*Spartium junceum*) decorates the landscape of the maquis in spring with its yellow blossoms. The maquis covers the karst surface and protects shallow stony soils from water erosion, and it is a shelter and nurturing place for numerous birds, insects, and reptiles. The maquis is a consequence of destruction of forest cover in the past decades, and frequent forest fires that prevent young trees from maturing.

Garrigue

Too frequent cutting of the underbrush, grazing, and similar negative effects due to soil erosion lead to development of the underbrush garrigue, a sparse scrub that does not exceed 1 m in height (Photo 4.24). Because of light penetration in the bushes other species grow that are present in the succession of oak forest. These are heliophilic and thermophilic species, such as rock rose (*Cistus incanus, C. creticus*, *C. salviifolius*, and *C. monspeliensis*).

In garrigue there are also aromatic plants that produce essential oil, like: the curry plant (*Helichrisum italicum*), thyme (*Tymus vulgaris*), salvia (*Salvia officinalis*), rosemary (*Rosmarinus officinalis*), tree heather (*Erica arborea*), Spanish broom (*Spartium junceum*), and thorned calicotome (*Calicotome villosa*). In garrigue it is possible to find plants of secondary importance such as *Pistacia lentiscus*, *Myrtus communis*, *Juniperus macrocarpa*, *Arbutus unedo*, *Lonicera implexa, Ruscus aculeatus, Olea europaea, Brachypodium*

Photo 4.23 Typical macquis with Christ's thorn (*Paliurus spina-Christi*) as the dominant variety—in the hinterland of the city of Zadar

Photo 4.24 Garrigue (yellow colored, *right* from vineyard) with Spanish broom (*Spartium junceum*) as the dominant plant species

retusum, and others. As a component of bushes there are *Carpinus orientalis*, *Acer monspessulanum*, and *Coronilla emeroides*. Because of its very dry state and high density, fires are common in the garrigue.

Rocky vegetation (Photo 4.25)

The Rocky vegetation area is the ultimate stage of degradation of forest holm oak (*Quercus ilex*), formed after strong wind and water erosion and after the destruction of

Photo 4.25 Mediterranean pasture with rocky vegetation

forests due to intensive grazing or fire. Huge areas have developed on the islands of Pag, Kornati, Goli, Prvić, and others. Developed soil is extremely shallow and skeletal, with lots of movable and immovable rocks. Rocky vegetation areas are used as pastures.

Upon cessation of grazing in the rocky overgrown areas of a garrigue, which is increasingly common with age, forest species begin to emerge. The most common plant species in rocky vegetation areas are: sage (*Salvia officinalis*), immortelle (*Helichrysum italicum*), *Drypis spinosa, Teucrium polium,* coastal heath (*Satureja montana*), and feather (*Stipa eriocaulis*). All of these species have a large amount of essential oils, often have thorns, or are covered with hair, such as browsing thorn used for the protection of domestic animals or wildlife.

A prolonged summer drought leads to an almost complete interruption in the vegetation of the rocky vegetation area, which is reinstated after the first autumn rains. Because of the mild winters of the Adriatic region, growth takes place in winter, and continues very intensively in the spring, when many surface rocks look like flowers mainly through bloom species of the genus *Cistus*. Species are richer in sub-Mediterranean rock areas, where there is more soil. Rock gardens make good bee pasture; species of the genus *Salvia* and *Erica* are especially valued honey plants.

4.3.4 Soil Fauna: Zoosphere of Soil

The versatile organisms or soil fauna living in soil remain there for their lifetime (pedofauna) or for just one phase of life. Each region has a characteristic soil fauna. For example, rodent squirrels live in the eastern Pannonian subregion, where steppe conditions prevail. They live in the soil and dig long tunnel systems, mixing the soil horizons. The squirrel is now rare, but traces of their life in the soil of the steppe area remain as krotovine—a section of corridor filled with the dark color of humus from A horizon (Photo 4.26)—seen in the yellow loess.

Besides squirrels, field mice and hamsters also live in the soil, living in a system of numerous holes that increase the infiltration of water leading to receipt of a large mass of water during the torrential rain which in summer is a regular occurrence. The most studied and most widespread and important member of the pedofauna is the earthworm (*Lumbricidae*). They do not favor the soils of the steppe zone because of the regular summer drought. Worms are a very important factor in soil formation, and an indicator of favorable conditions in the soil, because they live in fertile soils. They are like some microfauna (*Colembola, Acarina*), organisms that are very sensitive to all chemical contamination, and are used as an indicator of the presence of petrochemicals in the soil. Earthworms are most active in a loamy soil texture, with neutral chemistry, plenty of humus, and good wetness. If these conditions are not fulfilled, they pass into a latent state with the absence of any activity. Through their digestive system they mix the mass of soil forming structural aggregates, creating aggregates of optimal size and stability. Modern soil management should create and maintain conditions favorable for earthworms. Fertilizing with compost helps their propagation in the soil, only of course if other favorable conditions for them are met.

In humid areas, in desert ground the mole has a similar role to play as the squirrel (Photo 4.27). It lives in the soil, and in the surface layer eats macrofauna, mostly worms, and participates in the formation of a stable soil structure, soil mixing, and aeration.

Microorganisms have a tremendous importance in soil, considering that they are the main factor in the decomposition of dead organic matter, humification, and cycling of

Photo 4.26 Steppe rodent squirrels (*Spermophilus citellus*) (*left*); a network of corridors dug in the soil in the shallow profile of chernozem leaves dark circular formations—krotovine (*right*)

Photo 4.27 Mole-hills: a clear trail left by the mole on the crop (*left*) and in the meadow (*right*)

matter and energy in nature. We class them as microflora and microfauna. The most important representatives of the microflora are soil bacteria, which are extremely important factors of mineralization and humification, and other processes such as oxidation and reduction in soil, nitrogen fixation, and others. In Croatia, intense microbial activity has been investigated, with particular attention paid to the beneficial microorganisms—symbiotic and nonsymbiotic and nitrogen fixators. The results show the very effective application of bacterization (inoculation) of legume crops, especially soybeans (Redžepović et al. 2007, 2008; Huić-Babic et al. 2008).

4.3.5 Anthropogenic Influence on Soil Formation

Until the dawn of man and land use for growing plants in agriculture, soil formation took place under the exclusive influence of natural factors of soil genesis. Today, there is virtually no soil on Earth without human influence. Anthropogenic influence is a decisive factor in soil formation and soil evolution. Over the territory of Croatia, this influence has lasted from 5,000 years ago in the Vučedol culture (according to the Vučedol locality near Vukovar). Namely, when the first cities of Mesopotamia gave birth to the first town-states and Egypt was an early empire, Europe was lagging behind, at least, as previously thought. It was on the banks of the Danube, in Vučedol, where equally advanced civilizations were created. According to archeologists the "leap in civilization" that occurred about 5,000 years ago can be measured and compared only with the twentieth century. The people of this culture lived from agriculture, mostly beef cattle. The influence of man on the soil of this region has lasted from these times until today, i.e., 5,000 years. The types and intensity of the influence in some phases of the development of mankind and civilization were variable. Whether in organized agriculture or forestry production, the actions of man and their effect on soil genesis and the properties of soil had one clear and predetermined goal, the maximal yield of biomass. The degree of anthropogenic man's influence grows with demographic expansion; the creation of new urban areas is increasing the level of development, particularly the development of science and industry. Conscious actions of man can slow down, speed up or stop the natural flow of evolution and redirect soil genesis to obtain a changed and entirely new soil that does not exist in nature. In the Adriatic region, the soil would not survive without protection from loss by erosion. All impacts of humankind on soil are collectively called anthropogenization. No soil genesis factor can bring about such fast and intensive changes in natural ecosystems as man. The effects of anthropogenic influence on soil genesis can be positive or negative.

Through the conquest of a new agricultural land area at the expense of rangelands, man interrupts the natural flow of matter and directs it in another direction. All soils in steppes in Croatia are cultivated and natural steppe vegetation has been completely removed and is practically lost. Present and possibly future generations living on steppes will never see the characteristic vegetation and animals of steppe lands. Soil tillage, if not carried out along with other activities like fertilization and humus enrichment, as a rule leads to a decrease in humus content and some mineral elements in soil, because of the intensifying aerobic microbial processes of mineralization of organic matter. Taking the crop yield, man removes the soil's nutrients, reduces its content in soil, and decreases the natural fertility of soil. The use of heavy machinery causes soil compaction, and traffic on wet soil will result in destruction of soil. The application of some fertilizers causes soil acidification (Butorac and Bašić 1989).

In modern agriculture, economic factors place strong pressure on reducing crops or even applying a monoculture. This creates an unstable, artificial, sensitive agroecosystem, the soil becomes exhausted, and its fertility is gradually reduced.

A particular problem of the modern agricultural production and farming system are plant residues (straw, corn stalks, etc.). The economics of the classical procedures of their use are not justified—crop residues are used as animal feed in livestock production, where it is then returned as farmyard manure. Burning leads to a complete loss of nutrients contained in the residue. If bioecological conditions allow their decomposition, the plowing of plant residues is reasonable with appropriate interventions in terms of C:N ratio by nitrogen fertilization to control the quality of humification in the soil (Butorac 1999).

The destruction of natural forest vegetation, causing disruption of the water regime, raises the groundwater level, paving the way for water and wind erosion, floods, etc. The research results of Martinovic (1997) show that the Plitvicka jezera soil in arable land is much more susceptible to erosion than soils under forests, with reduced chemistry, humus content, and stability of structural aggregates, and in the interest of protecting the National Park Plitvicka jezera, reduction of agriculture to the benefit of forest vegetation is recommended.

4.3.6 Relief

Unlike other factors of soil genesis acting directly, relief affects soil formation indirectly, by changing and correcting other factors, climate conditions in the first place. Relief is defined as the shape of the earth's surface, characterized by horizontal and vertical dimensions. The exposition and inclination of the slope influences soil genesis by redistributing matter and energy—solar heat, rainwater, and correcting the climate as a factor in the genesis of the soil. It affects the heating of the soil, and gravity and rainwater moves soil via runoff and erosion transport and sediments it in stable positions at the bottom of slopes or in the plain. Relief has an influence on the distribution of plant species: for example thermophilic ones prefer south-oriented slopes, and sloping terrain makes possible the occurrence of avalanches and landslides and specific vegetation that accompanies such sites, like wild climber (*Clematis vitalba*).

Flat forms of relief accept the entire water mass from rainfall. From higher topographic positions through runoff the water suspension flows down the slope with suspended soil material, and in relief forms like depressions or plains the soil accumulates as a multilayer sediment. Inclination, especially of tilled soils, causes erosion and slide processes returning soil genesis to the early, juvenile phases (Photo

Photo 4.28 The initial slide of a soil mass is an process to be expected on the slope of Kalnik hill after deep tillage for a vine plantation without proper soil conservation

Photo 4.29 Toposequence of soils of Vukovar loess plateau by Ovčara visible on freshly plowed soil

4.28). The removal of mineral and organic soil particles by erosion on sloping terrain by severe water flow interferes with the development of soil on the surface and stops its evolution. With complete removal of soil erosion, soil can stay in the initial stages of evolution. Conversely, the erosion-affected soil and translocated material is deposited on the lower slopes as colluvium on the existing soil cover, the genesis of their previous termination, and it may represent a new parent rock (regolith) for the initial stages of the genesis of a new soil.

In a similar way, relief as an important factor affects the hydrothermal regime by redistribution of solar energy. The

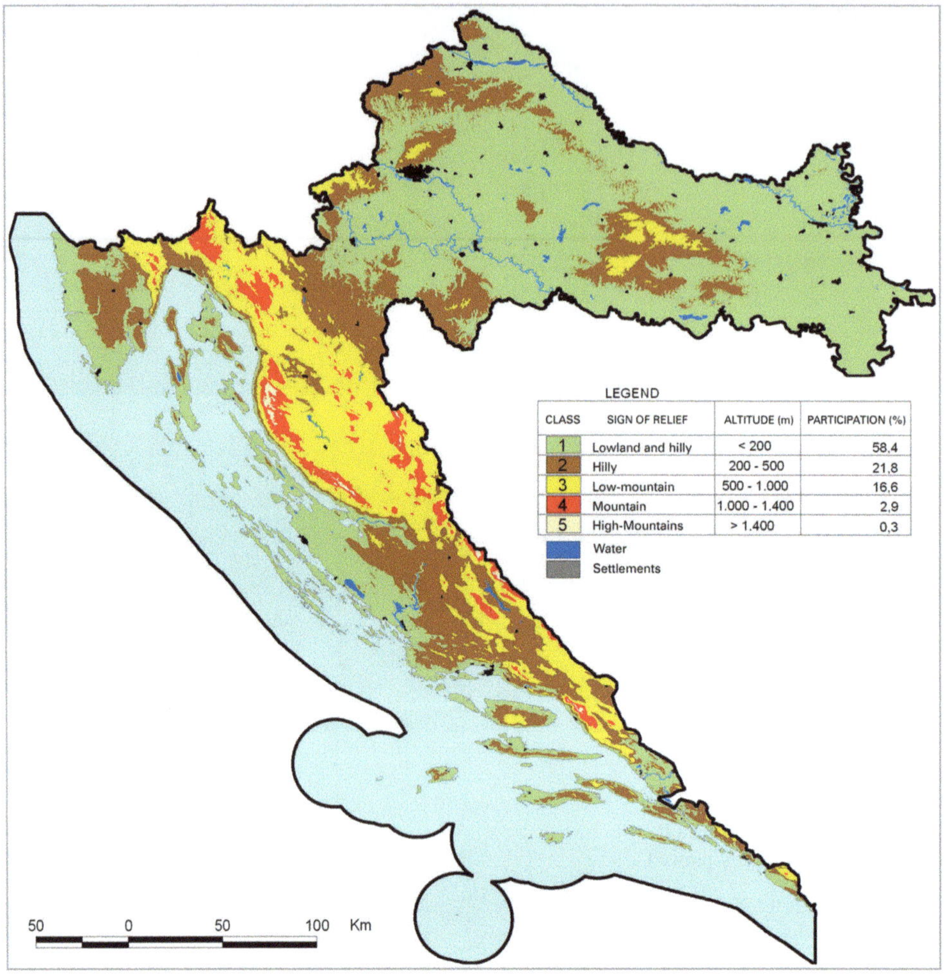

Fig. 4.5 Altitude above sea level of different forms of relief in the territory of Croatia (Husnjak 2000)

adoption of solar energy is affected by the slope exposure (sun exposure) and inclination. The strongest warming is when the sun's rays fall on the soil surface at an angle of 90 °C. Therefore, the warmest are southern slopes, followed by west-, then east- and finally north-oriented slopes.

As a factor in redistribution of water and heat, relief influences the hydrothermic regime contributing to climate and solar correction of the pedoclimate. Various thermal and wetting conditions lead to the occurrence of different natural plant species. Relief is an important factor in soil genesis, which together with the effect of other factors directs the process of soil formation. Some slow it down, some speed it up and/or stop the process, causing changes to different soil units in an area or a known soil toposequence (Photo 4.29). Thus, in one small area soil can be found in the initial phase of soil evolution while another close by can be in a developed phase—i.e., young and old phases may be very close to one another. The photo shows an example in the loess on the Vukovar loess plateau.

The yellowish-white color is regosol on loess generated on "top" of ploughed loess dunes by water erosion; on the slope of dunes we find cambisol eutric on loess, but on the bottom an initial phase of luvisol. The dominant factor is the influence of forms of relief on the redistribution of precipitation. The most important indicator that reflects the opportunities presented by relief of an area is the altitude of an area. Figure 4.5 presents the altitude of the Croatian territory, divided into five classes of relief.

As can be seen, based on relief the land is dominated by an altitude up to 200 m a.s.l., i.e., lowland and hillock relief with 58 % of the total area of Croatia, followed by a hilly relief with a.s.l. 200–500 m, which occupies about 22 %, then the mountain relief with altitudes of 500–1,000 m, which occupies 17 % of Croatian territory. The low mountain relief (1,000–1,400 m a.s.l.) has only 3 %, while the smallest (only 0.3 % of the area) is represented by the high mountain relief with altitudes over 1,400 m a.s.l.

In the whole eastern Pannonian subregion the plain is the dominant relief form with very fertile soils—chernozem typically, and humogley (Photos 4.30 and 4.31).

An important indicator of the opportunity of relief to illustrate the conditions of origin and translocation of soil is

Photo 4.30 Panorama of the "endless" plain of west Syrmium with chernozem on loess visible from Principovac—the famous belvedere of Ilok, west Syrmium (P-1 subregion). In front the famous Ilok vineyards rise on the hillside of Fruška gora on cambisol eutric on loess

steepness, which is shown in Fig. 4.6, where the classification of slope is made according to criteria adopted by the International Geographical Transferring Company (Husnjak 2000). According to this classification, the slope is classified into six classes.

As can be seen, with 52 % the most widespread is the first class, the plains with slope of terrain < 3.5 %, where mass movement is not observed. The other classes of slope of terrain vary from gentle to bold to strong mass-moving surface runoff and erosion, and destructive capabilities. In other words, as much as 48 % of Croatia has conditions that are conducive to water erosion, and demolition of loose soil and parent material.

4.3.7 Time (Soil Age) as a Factor of Soil Genesis

Time is an important but passive factor in soil genesis and the time-consuming process of soil formation. Most Croatian soils have been created since the last glaciations, i.e., within a period of about 10,000 years. It is known that during this period we have had significant changes, for example climatic conditions have changed, and were not as they are today. But as much as 48 % of Croatian territory is exposed to surface runoff of different intensities, and it is clear that the surface area of all soils created by soil genesis may have their upper or lower layers translocated to lower positions, where they remain on slopes, at their foot, or are exposed by a watercourse to alluvial transfer.

Given the time and conditions under which they are made, soil can be classified as monogenetic soils, which were generated under constant factors of soil genesis, and polygenetic soils, which evolved over a longer period, in

Photo 4.31 A large plain with chernozem—Klisa near Vukovar. The highest "hill" in the landscape is a stack of sugar beet roots in autumn. In this region there is a joke in the circle of agronomists—the only task of the expert on this soil is not to forget to sow—the rest is the worry of nature (the Almighty)

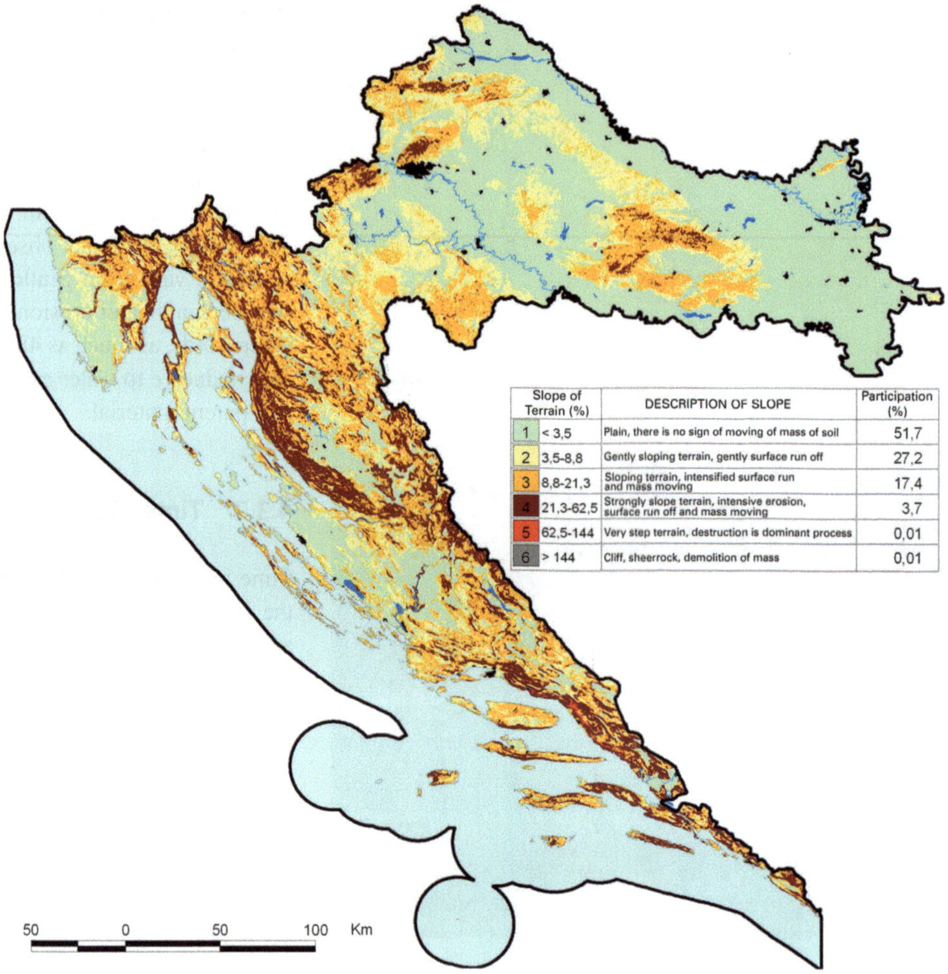

Fig. 4.6 Map of slope of Croatia (Husnjak 2000)

which the factors of soil genesis changed and certainly were different from what they are today. Polygenetic soils are prevalent, which makes a reconstruction of the character and endomorphological markers of these processes necessary in the soil profile. Our oldest soils are red soils on limestone, particularly the relict terra rossa, which goes back to the Tertiary.

To estimate the age of the soil we rely on the morphology of the profile—the number, characteristics, and especially the thickness and sequence of genetic horizons. This approach is reliable only if the parent material from the beginning of genesis was the same without the influence of impurities. However, climatic, geomorphological, and lithological relationships and circumstances have changed, so it is very difficult to reconstruct according to these criteria. Reliable determination of the age of soil organic matter with ^{14}C in different soil horizons is performed and the age is determined as an average age of all horizons. In any case, the duration of soil genesis is highly variable, from very fast if a few cm of soil is created over 100 years as is the case with volcanic ash under tropical conditions and very slow 1 cm in 5,000 years on chalk in cold climate conditions. Theoretically, every year should create at least as much soil as is lost by erosion.

Monogenic soil is defined by some recent classifications as soil that arises and evolves in accordance with the present state of the factors and processes of soil genesis. This development may be in the initial stages of evolution, i.e., a young soil, or in the later stages of evolution, i.e., more developed, or there is development of a fully completed profile in accordance with these factors, which means that they are in dynamic equilibrium with the actual factors of soil genesis—the climax stage of development.

Unlike monogenetic soils, polygenetic soils may be relict and paleo soils. Relict soils may show the features of previous (different from today's) combinations of factors and processes, and their features, or the constructs that they contain, remain as relicts deep in the profile, but surface layers continue with soil evolution in accordance with recent conditions of soil genesis and evolution. Such relict soils can be maintained in equilibrium or are slower to change, and their origin may be the result of geological and climatic relicts of the past. For example, from the early Holocene (e.g., our chernozem of Boreal age), from the

Pleistocene (e.g., marbled, similar to the tundra soil with gray-red smearing and concretions originating from interglacial periods), or from the pre-Pleistocene period. An example of such a soil is the ferralitic soils of tropical and subtropical areas and our red soil—terra rossa.

Unlike relict soils, paleosoils are old soil covered with younger sediments of Pleistocene age, which partially or completely interrupted soil evolution. The evolution is totally "broken" in fossil soils, which are covered by younger sediments or are found deep in the cracks of karst. In the premountainous subregion (G-1) we find the so-called two-layer soil profiles, where the deeper layer is a result of old soil genesis but is covered by a shallow loess-like loam of eolian origin which is included in recent processes of soil genesis. In this loess area, we also see the emergence of the paleosoil to the surface because other layers have been removed by erosion. This soil is very similar to the modern morphology of cambisol eutric, but it is due to the mixing of calcareous loess and is significantly different from the "real" recent cambisol eutric on loess.

References

Bašić F (1976) Pedologija, Viša poljoprivredna škola Križevci, udžbenik, str. 231, Križevci

Bašić F (1982) Pedologija, II. preuređeno i dopunjeno izdanje, str. 184, Križevci

Bogunović M, Pavlić V, Bašić F, Šalinović I, Bašić I (1984) Pedološka karta dijela sekcije Vukovar-1,2 i 4, i Bačka Palanka-3, posebno izdanje Pro-jektnog savjeta za izradu OPK Hrvatske, tisak VGI, Beograd

Bogunović M, Vidaček Ž, Racz Z, Husnjak S, Sraka M (1996) Namjenska pedološka karta RH, mjerilo 1:300 000, Agronomski fakultet Sveučilišta u Zagrebu, Zavod za pedologiju, str. 46, Zagreb

Bašić F, Bašić I (2007) Navodnjavanjem tala suzbiti opustinjavanje Iloka!, Gazophylacium, God. XII br. 1-2 str. 73–93, Zagreb

Bašić F, Herceg (2010) Temelji uzgoja bilja, udžbenik, Sveučilište u Mostaru, tisak „Grafotisak" - Grude, str. 454, Mostar

Butorac A, Bašić F (1989) Antropogena zbijanja tla i mogući putovi njihovog uklanjanja ili ublažavanja. – Znanstveni skup posvećen 100. obljetnici rođenja Prof. dr. Dobroslava Todorovića - Zemun, posebno izdanje, Naučna knjiga – Beograd

Butorac A (1999) Opća agronomija, Sveučilišni udžbenik. Školska knjiga, str. 648, Zagreb

Durn G (2003) Terra Rossa in the Mediterranean Region: Parent Materials, Composition and Origin, Geologia Croatica: journal of the Croatian Geological Survey and the Croatian Geological Society. 56/1 1; p 83–100, Zagreb

Gajić-Čapka M (2004), Klima, Nacionalni program ublažavanja posljedica suša i suzbijanja oštećenja zemljišta (NAP), Državni hidrometeorološki zavod, str. 70, Zagreb

Galović L, Frechen M, Halamić J, Durn G, Romić M (2009) Loess Chronostratigraphy in Eastern Croatia, A Luminescence Dating Approach, Quaternary international, Elsevier,

Gračanin M (1950) Mjesečni kišni faktori i njihovo značenje u pedološkim istraživanjima, Poljoprivredna znanstvena smotra, Sv. 12, p. 51-66, Zagreb

Gračanin M (1951) Pedologija – III dio – Sistematika tala, udžbenik Sveučilišta u Zagrebu, Školska knjiga, str. 298, Zagreb

Huić Babić K, Schauss K, Hai B, Sikora S, Redžepović S, Radl V, Schloter M (2008) Influence of different Sinorhizobium meliloti inocula on abundance of genes involved in nitrogen transformations in the rhizosphere of alfalfa (Medicago sativa L.), Environmental Microbiology, 10 11: 2922–2930

Husnjak S (2000) Procjena rizika erozije tla vodom metodom kartiranja u Hrvatskoj. Disertacija, Agronomski fakultet Sveučilištau u Zagrebu, Zagreb

Husnjak S (2012) Sistematika tala Hrvatske. Zavod za pedologiju, Sveučilište u Zagrebu, Agronomski fakultet, Zagreb

Kustura A, Kisić I, Bašić F, Jurišić A (2008) Soil Erosion by Water in Perennial Plantations of the Ilok Region, Agriculturae Conspectus Scientificus 73/2: p. 75–83

Kišpatić M (1877) Zemljoznanstvo obzirom na šumarstvo i gospodarstvo, troškom Kraljevske hrvatske zemaljske vlade, Zagreb

Martinović J (1997) Tloznanstvo u zaštiti okoliša: priručnik za inženjere, Državna uprava za zaštitu okoliša, str. 221, Zagreb

Redžepović S, Sikora S, Čolo J, Blažinkov M, Pecina M (2007) Influence of plant growth regulator and rhizobial inoculation on nodulation and soybean nitrogen content, Cereal Research Communication, 35, 2 Part 2; p 993–996

Redžepović S, Čolo J, Blažinkov M, Sikora S, Pecina M, Duraković L (2008) Utjecaj biostimulatora rasta i fungicida za tretiranje sjemena soje na učinkovitost simbiozne fiksacije dušika. Sjemenarstvo 24:169–176

Resulović H, Ćustović H (2002) Pedologija, opći dio, (knjiga I). Univerzitetski udžbenik, Sarajevo

Škorić A i sur (1977) Tla Slavonije i Baranje. Projektni savjet za izradu Pedološke karte Republike Hrvatske, str. 256, Zagreb

Škorić A i sur (1987) Pedosfera Istre, Projektni savjet za izradu pedološke karte Republike Hrvatske, str. 92, Zagreb

Škorić A i sur (2003) Tla gorske Hrvatske, monografija, Ministarstvo zaštite okoliša i prostornog uređenjag fakulteta, str. 169, Zagreb

Špoljar A, Kušec V, Kamenjak D, Kvaternjak I., Peremin-Volf T, (2006) Promjene pedoloških značajki đurđevačkih pijesaka uzrokovane revitalizacijom, Agronomski glasnik br. 3, str. 181–198

Vrbek B, Pernar N (2001) Status and prospects of soil science development in Croatia, Plenary paper, Summaries of IX Congress of Croatian Society of Soil Science, p 5–17. Brijuni

Vukelić J, Rauš Đ (1998) Šumarska fitocenologija, udžbenik Sveučilišta u Zagrebu, str. 310, Zagreb

Vukelić J, Mikac S, Baričević D, Bakšić D, Rosavec R (2008) Šumska staništa i šumske zajednice u Hrvatskoj, nacionalna ekološka mreža, monografija, Državni zavod za zaštitu prirode, st. 263, Zagreb

Zaninović K, Gajić-Čapka M, Perčec-Tadić M, et al (2008) Klimatski atlas Hrvatske, Climate Atlas of Croatia 1961–1990, 1971–2000, Državni hidrometeorološki zavod, str 200, Zagreb

Conferences and Other Sources

Towards environmental pressure indicators for the EU, European Commission and Eurostat (1999) Theme 8-Environment and energy

World reference base for soil resources 2006 (2006) World soil resources reports 103, Food and Agriculture Organization of The United Nations, p 128, Rome

5 Soil Taxonomy of Croatia

5.1 Short History of Soil Taxonomy

The classification of subjects of research is a key issue in any science, and soil science in this regard is no exception. But while all types of plants, animals, rocks, etc. are simply *identifiable*, with soils this is not the case, because soil is by definition a natural body in constant change and development—the evolution and taxonomy of taxons (soil units) means that we face the determination of the stage of evolution the actual taxon finds itself in at the moment of our observations. The lack of a unique, globally usable soil taxonomy system represented, and still represents, a source of controversy and barrier to the development of modern soil science. This was the main motive behind key developments in this field, first created by the FAO/UNESCO, and then the laudable WRB soil taxonomy (WRB 2006). The factors affecting recent soil formation and evolution are variable and certainly different from those that influenced past soils. This is particularly true of polygenetic soils. For example, red soil (terra rossa) in the Mountain region (G-2 subregion) of Croatia was created under the conditions of the hot and humid climate of the Tertiary, which is wholly inconsistent with the current climate. Science-based soil taxonomy requires a clear definition of taxons—units of taxonomy and a systematic approach that leaves "empty place-holders" for taxons that have not yet been determined. There is a wealth of experience regarding taxonomy of the soils of Croatia. As in other, more advanced countries at the time, the first data on soils were collected within the related natural sciences, mostly geological sciences, but in the late seventeenth century soil science became dominated by the influence of the western European, especially German schools. Progress was made in the development of soil science in the mid-nineteenth century through the establishment of the Higher Royal Agricultural and Forestry School in Križevci, in which arose the need for new knowledge and a systematic arrangement of the collected empirical knowledge on soil. Faced with this situation, the first professor of Soil Science, Mijo Kispatic, published the first textbook on soil in the Croatian language and this was among the first such books in the world. It was entitled *Zemljoznanstvo*, and was issued in 1877. A special chapter, entitled "judgment of soil", briefly discusses the possibility of soil taxonomy according to natural history and economic principles. For example, the taxonomy of soils based on the principles of the natural sciences is represented by taxonomy based on geological origin and the (physical and chemical) properties of the soil. Taxonomy based on economic principles intends to classify soils according to suitability for crop production in agriculture, represented in the book by an Austrian soil taxonomy with a classification according to suitability for growing wheat, barley, oats, and rye.

Francis Sandor, the first head of the National Institute for Soil Research, established by the decision of the Royal Croatian–Slavonian–Dalmatian Government in 1910 in Zagreb, published a book in 1914 entitled *Soil Science I—General Soil Science*. The textbook contains a chapter on "Soil taxonomy," stating that soil taxonomy can be based on scientific and empirical principles. It is very important to note that Sandor focused and directed research in Croatian soil science towards the Russian "genetic school" of soil science, as did the whole of Europe at that time. Sandor presents the taxonomy by Glinka, who systematized the soil into two groups: ectodynamomorphic soils, dominated by the influence of external factors, and endodynamomorphic soils, where the dominant influences are the internal properties and the parent material. This taxonomy was retained in Croatian soil science until World War II. On this period it is important to note that at the international conference in Stockholm in 1911 a "Sketch of climazonal soils of the Kingdom of Croatia and Slavonia" was shown, which was created by Gorijanović-Kramberger and Sandor (Šandor 1911). This confirms a long tradition of soil taxonomy in Croatia, which is permanently attached to the "genetic principles."

The biggest contribution to the overall development of soil science in Croatia to date is certainly that of Michael Gračanin, a professor in the Agriculture and Forestry Faculty in Zagreb. His textbook *Pedology—Part III—Systematics of*

F. Bašić, *The Soils of Croatia*, World Soils Book Series,
DOI: 10.1007/978-94-007-5815-5_5,

Table 5.1 Systematic units of soil taxonomy of typical soils (Gračanin 1951)

Taxon name	Criteria of taxonomy
Class	The dynamics of the mobile components of soil (water and air)
Orders	Characteristics of soil type-generating processes
Types	Characteristics of soil type-generating processes and physiographic features of soil type
Subtypes	Stage of soil evolution
Varieties	The development and expression of genetic horizons and subhorizons
Subvarieties	Expression of specific genetic characters and features
Species	Texture of soil profile
Subspecies	The mineralogical and chemical features of coarse fraction of soil
Facies	Geological characteristics of the parent substrate

Soil (Gračanin 1951) presented the soil taxonomy system known at the time, which was divided into geological–petrographic, physical, chemical, morphological, and genetic taxonomy. He divided all soils of the pedosphere into divisions: developed or typical soils and undeveloped or atypical soils. Developed or typical soils have physiographic features in accordance with soil "type-generating processes." On this basis he built the soil taxonomy based on genetic principles. The central unit of this soil taxonomy is soil type. The taxonomy relied on the properties of soil type and recent soil "type-generating pedogenetic processes." "Type of soil" implies a systematic unit that is characterized by the soil type's physiographic characteristics and soil type-generating processes. Units of soil genesis (taxons) are shown in Table 5.1.

The order of undeveloped, atypical soils includes soils that do not have expressed physiographic features in accordance with standard pedogenetic processes in the areas in which they are reported (Table 5.2).

The soil taxonomy of Gračanin and his term "type-generating soil genesis process" have been accepted and used for a long time in Croatian soil science. Each soil type in this system accurately describes the soil type-generating processes by which they arise and the conditions under which it was possible. Similarly, the continuous use of the term "climax stage of the soil" represents the stage at which the soil is fully compliant with the current climate and other conditions in which we find a particular type of soil. He also established the stages of soil evolution, and thus formed a very logical system in which there are no "empty spaces." In addition, it is important to emphasize that some terms and criteria were lifted from each category in this taxonomy and used by pedologists in Croatia and abroad in their own soil taxonomy systems. Looking back over a time period of more than half a century, we believe that the deviation from those principles was the wrong way forward for Croatian soil taxonomy, as it did not adhere to scientific principles.

After the middle of the last century, the development of soil taxonomy in Croatia continued within the Yugoslav Society of Soil Science (YuSSS). A significant event in this period was the first proposal of a taxonomy of the soils of Yugoslavia, whose authors were Croatian pedologists Neugebauer and Skoric, and Ciric, Filipovski and Zivkovic from other republics (now countries) of the former Yugoslavia. The proposal was accepted at the second YuSSS Congress in Ohrid in 1963 (Neugebauer et al. 1963). The taxonomy was based on genetic principles, and the systematic units of soil taxonomy were:

- soil orders—based on origin and intensity of soil wetting,
- class of soil—based on number and sequence of genetic horizons in the soil profile,
- soil type—defined as the central unit of taxonomy.

The lower systematic units (subtype, variety, and form) were defined by criteria that are most different in relation to soil type fertility, as well as its most important feature.

Although this taxonomy was retained for only 10 years, it had great importance for the further development of pedology in general and in particular soil taxonomy in the former Yugoslavia. It was a period of great progress in soil science because all the republics as well as autonomous

Table 5.2 Systematic units of taxonomy of atypical soils (Gračanin 1951)

Taxon name	Criteria of systematization on lower soil units
Class	Characteristics of mean soil genesis processes
Order	Chemical properties of soil
Genus	Petrographic characteristics of parent material detritus
Subgenus	Geological origin of parent material
Variety	The depth of soil profile
Species	The degree of fragmentation of regolith

Table 5.3 Systematic units of soil taxonomy of YuSSS (Škorić et al. 1973)

Soil systematic unit	Criteria of definition of systematic units
Soil division	Based on the character of soil wetting and composition of the water, there are four divisions: automorphic, hydromorphic, halomorphic, and subaquatic soils.
Class of soil	Group of soils of uniform genetic horizons, which represent stage of soil evolution
Soil type	Uniform sequence of genetic horizons, uniform basic processes of transformation and migration of mineral and organic matter, similar physical and chemical characteristics of genetic horizons.
Subtype, variety, and form	Criteria for systematic units are different for each soil type

provinces (Kosovo and Vojvodina) established institutes or universities with a soil science unit. The system of soil taxonomy was for the first time designed to provide users with specific information regarding the production potential of the soil, which in those days was considered particularly valuable. But, as in other European taxonomies of that time, this taxonomy totally ignored all the other functions, including the already known ecological, regulatory functions of soil. It is important to remember that at that time intensive agriculture and/or forestry were the unquestioned focus of development and maximal yield (of crop or timber mass) per unit area was in practice the only criteria of efficiency of agriculture and forestry.

The next contribution to the development of soil taxonomy in Croatia was the creation of a detailed soil classification of Croatian Posavina, presented at the third YuSSS Congress held in Zadar in 1973. With partial reaffirmation of the taxonomic principles of Gračanin for the systematization of the lower units of soil type, the authors proposed uniform criteria and concepts introduced in U.S. soil taxonomy such as series, type, and phase of the soil. The more detailed taxonomy of the lower units of soil types according to this system is natural, transparent, and predicts the existence of systematic units with specific soil properties and possibly "empty places." However, this taxonomy has remained without significant impact. It follows the creation of the *Soil Taxonomy of Yugoslavia* by a group of authors, the first of which was Croatian soil scientist Skoric. The taxonomic proposal was presented to the fourth YuSSS Congress in 1972 in Belgrade. The classification was accepted, corrected, and supplemented by the YuSSS Commission on soil genesis, taxonomy, and mapping of soils and printed as a special edition that became the official *Soil Taxonomy of YuSSS* (Škorić et al. 1973). This taxonomy represents a further development of the previous soil taxonomy system, with considerable extra work founded on the same basis. The most important features of this taxonomy are that it is founded on the principles of the "genetic school", and the subordination of production potential of soils into production of organic matter. The central unit of this taxonomy is also soil type. At a level of the taxonomy "higher" than *type* genetic criteria are used, but lower levels use criteria that indicate the production potential of soils. The classification is based on clearly defined soil properties that are morphologically visible and measurable. The basis of this taxonomy is shown in Table 5.3.

This taxonomy was used in the former Yugoslavia, and especially important was its use for the systematization of soils in the General Soil Map of Croatia at the scale of 1:50,000, which was an opportunity to test in practice the applicability of the new taxonomy. It should be noted that the documentation for the preparation of soil maps and printed sheets of the map were labeled "state secret" and therefore were excluded from the exchange of opinions and critical judgments of other professions and colleagues within the country and especially abroad. This development in isolation does not enrich the fund of knowledge, as the work is kept within a closed circle. The taxonomy does not recognize any function of soil other than the production of organic matter in agriculture and forestry. Soil degradation and soil-protection problems have no place in the taxonomy. No attention was given to this problem within the epochal project GSM 1:50,000. However, after a 10-year period of intense research and use of this taxonomy, the increased level of knowledge and new information collected on soils were used to raise suggestions for its amendment, among which were no proposals to radically change the basic principles on which it was based. The same authors therefore prepared a revised version called the *Soil Taxonomy of Yugoslavia* (Škorić et al. 1985). Accepted proposals were related to a number of amendments, definitions, and designations of horizons and subhorizons, along with amendments to the divisions and classes and their definitions, as well as amendments to soil types, lower units, and their definitions.

Work on Croatian soil taxonomy over the last two decades has largely related to the use of international classifications defined by the FAO/UNESCO and WRB (World Reference Base) taxonomy systems. Croatia has been involved in the project MARS (Monitoring Agriculture with Remote Sensing), whose goal was a soil map of Europe at the scale of 1:1,000,000 based on FAO/UNESCO taxonomy. This taxonomy was created in 1974 when it was developed for the legend of the Soil Map of the World as a unified soil taxonomy system. For the purposes of the MARS project, the FAO/UNESCO initiative led to the Soil Map of Croatia, 1:1,000,000, which was printed as a separate edition with an associated soil database, and the legend was prepared as a correlation between the soil systematic units of the current Croatian soil taxonomy and the FAO/UNESCO taxonomy. Although the introduction of these principles of soil classification represented a scientific contribution to the development of soil taxonomy in Croatia, for various reasons it has not found wider application in the world or in Croatia (Špoljar 1999; Špoljar et al. 2006). These were the driving forces behind the initiative to develop an international soil taxonomy system that began in 1980 and was finished in 1998 when a draft version of the World Reference Base for soil (World Reference Base for Soil Resources—WRB) was presented. It was composed based on the FAO/UNESCO legend and a unique system of soil classification, which would be recognized by the International Society of Soil Science. The aim was to facilitate communication in the field of soil systematization through correlation with the national WRB classification. During the last decade, we have been witness to the activities of national societies as they develop correlations of national classifications with the WRB. In Croatia, too, a systematic investigation of the correlative relationship with the WRB was initiated (Racz 1999; Vrbek et al. 2008; Husnjak et al. 2010). It was the firm opinion of the circle of soil scientists and professionals behind Croatian taxonomy that it was necessary to introduce a number of principles, criteria, and laboratory methods, used in the international WRB taxonomy system, to soil taxonomy in Croatia.

5.2 Actual Soil Taxonomy

A modern soil taxonomy would include changes introduced by the WRB for Soil Resources (2006). Bearing this in mind, we designed and proposed a special research project oriented towards Croatian soil systematics. We believe that the most acceptable and most readable soil taxonomy is that created by Husnjak (2012), as leader of that project. As with the previous taxonomy, this is also based on a hierarchical system, which consists of the following hierarchical sequence:

Soil order **→** ***Class*** **→** ***TYPE*** **→** ***Subtype*** **→** ***Variety*** **→** ***Form***

The central unit is the soil type but the highest hierarchical unit is the *order*, based upon the form and intensity of soil wetting, hydromorphism.

5.2.1 Definitions of Genetic Soil Horizons

Genetic horizons are labeled with upper-case letters; a supplemental label specifying the nature of the horizon or subhorizons is in lower-case letters.

5.2.1.1 Horizons of Accumulation of Organic Matter

O—(Organic) is a surface horizon rich in organic matter in different stages of decomposition, lying on the mineral soil and formed under aerobic conditions. Various subhorizons are denoted according to the degree of decomposition of organic matter:

- O_l—(Litter) is a subhorizon of unaltered organic matter with the clearly perceived character of the organic residues and plant species from which they originated.
- O_f—(fermentation) is a subhorizon of half-decomposed organic substances where it is difficult to perceive the origin of the macroscopic organic residues.
- O_h—(humus), a humified organic subhorizon in contact with mineral soil, is reported to involve the mineral grains of soil.

(A)—(accumulation) initial humus-accumulative horizon, poorly developed; in fact, a biologically activated surface layer of soil or a layer of dark soil with intermittent discharge of source rocks to the surface.

A—Typical humus-accumulative horizon in which the humus is mixed with mineral soil. Subhorizons are denoted according to the different character of the humus:

- A_{mo}—(mollis—soft, bland) molic horizon, deep, dark brown, with a mild, base-saturated humus and prominent structure.
- A_{um}—(umbra—shades) umbric A horizon, as dark and deep as the molic one, but in contrast to it a massive structure in the dry state, and humus saturation by bases is below 50 %.
- A_{oh}—(ochros—pale) ochric A horizon, brighter colors and shallower than the molic and umbric horizon, a weakly expressed structure, and hard and compact when dry.
- A_{na}—(aquatic) is a variant of the A horizon resulting under hydromorphic conditions, bluish black or dark gray color with greases and concretions.

5.2.1.2 Cambic Horizons of Soils

According to their origin, there are two forms of cambic subhorizon:

- $(B)_o$—(o—Croatian: clay formation) is a cambic horizon originating from increased weathering of primary and formation of secondary minerals in the process of argilogenesis. As it contains iron oxide, which is in the process of being released, this horizon has a reddish brown color.

- $(B)_{no}$—(insoluble residue accumulation) accumulation of insoluble residue formed by dissolving $CaCO_3$ in limestone and dolomite.

5.2.1.3 Horizons of Eluvial-Illuvial Processes

E—(eluviation—flushing) eluvial horizon has a bright color and, in relation to the horizon below it, contains less clay, humus, or sesquioxide. In terms of the destruction of clay during the process of podzolization, it takes on the color of ash and is labeled a spodic (spodos—ash) E horizon.

B—illuvial horizon which lies below the eluvial one and, in relation to it, contains more clay, humus, or sesquioxide, giving the various forms of the illuvial horizon:

- B_t—(tone-clay) argiluvic subhorizon, arises in the process of clay illuviation. On the surface of structural aggregates and pores is a macroscopically visible membrane of illuvial clay particles.
- B_h—subhorizon of illuviation of humus in the process of podzolization; below the spodic E horizon it is brown to black.
- B_{fe}—(ferrous) subhorizon of reddish brown color from the sesquioxide illuviated after decomposition of clay minerals in the process of podzolization.

5.2.1.4 Horizons of Processes of Gleyization, Pseudogleyization, and Peat Genesis

G—(gley) gley horizon arises under the influence of excess water. Depending on the dominant process it is classified into different subhorizons:

- G_r—(reduction) horizon bluish or greenish color, in which water is permanently stagnant and dominated by processes of reduction.
- G_{so}—(secondary oxidation) horizon of rusty color, in which the oxidation processes dominate over reduction.

S—(stagnation of rainwater) compacting, impermeable horizon of stagnation of rainwater and alteration of wet and dry phases in the process of pseudogleyization, streaked with gray and rusty albeluvic tonguing.

T—(peat) horizon of accumulation of poorly decomposed organic matter in the water under anaerobic conditions.

5.2.1.5 Anthropogenic Horizons

P—(plow) horizon of anthropogenic treatment—deep plowing with mixing of multiple (at least two), genetic horizons of natural soils.

5.2.1.6 Parent Substrate

C—indicates a loose rock or parent substrate, with detritus and hard, massive rocks.

R—(rock) indicates a hard, massive, parent substrate.

The layers of multilayer soils characterized by deposited soils are indicated by Roman numerals I, II, III, etc.

The depth of each horizon is indicated in brackets marking the depth at which the horizon starts and finishes.

Depth of soil is considered to be the total depth from the surface to the unchanged parent substrate. It is more easily determined in soils on hard and harder than loose substrates. It depends on many factors: generally the deepest are soils of tropical areas, because the depth and intensity of wear is the greatest; shallower soils occur in temperate climates, and soils of boreal and polar climates are shallower still. Under the same climatic conditions, soils over loose substrates are deeper than those over hard, soils in depressions on slopes and hilltops are also deeper, as are older soils compared to younger soils.

The depth of soil has great ecological importance. With other conditions equal, deeper soils are more fertile because the shallow vegetation uses water and nutrients from a larger mass of soil. The physiologically active depth is the depth to which roots of the plant penetrate. It cannot be equated with the depth of soil, the solum. If the soil has a heavily compacted horizon as in acid soil, or groundwater is present, the physiologically active depth is less than the depth of the solum. The physiologically active depth is not identical with any plow layer depth, because plant roots (rhizosphere) penetrate to a depth greater than the depth of plowing, if there are water and nutrients in the deeper layer.

According to the depth of the soil solum in Croatian soil taxonomy, soils are divided into: very shallow (<10 cm), shallow (10–30 cm), medium deep (30–60 cm), deep (60–120 cm) and very deep (>120 cm).

To assess the fertility of soil, an important indicator is the depth of particular horizons. Under variable conditions, the same soil with a deeper A horizon is more fruitful than a shallow soil horizon and a soil with a deeper eluvial horizon is less fertile than soils with a shallow E horizon, etc.

As stated in the previous section, the origin of water and duration of sufficient wetting led to classification into five orders: Terrestrial (automorphic), Semiterrestrial (semihydromorphic), Hydromorphic, Halomorphic, and Subaquatic soils. The integral taxonomy adopted is presented in Table 5.4.

5.2.2 Soil Taxonomy

Soils with sufficient wetting and hydromorphic processes caused by water from rainfall are variously classified in the previous taxonomy of soils, some within the automorphic, and others within hydromorphic categories. Unlike all previous taxonomies, Husnjak (2012) distinguishes a special order of semiterrestrial soils. Therefore, a special place is given in the taxonomy to a soil under the influence of water from rainfall in some horizons of the profile where periodically alternating periods of stagnation of rainwater

Table 5.4 Soil taxonomy of Croatia (Husnjak 2012)

Type	Subtype
(A) Order of terrestrial soils	
Class I. Initial automorphic soils (A)–R or (A)–C soil horizon sequence	
Lithosol	Based on parent rock—origin of mineral component of soil
Regosol	
Arenosol	Formed on quartzite sand (>95 % quartz)
	Formed on siliceous sand (<95 % quartz)
	Formed on siliceous—calcareous (<95 % quartz)
Colluvial soil	Calcareous
	Eutric, siliceous
	Dystric, siliceous
	Colluvium on fossil soil
	Colluvial—fluvial
Class II. Humus-accumulative soils A–C, A–R, or A–C–R soil horizon sequence	
Kalkomelanosol	Typical
	Cambic
	Cambisol rhodic (terra rossa)
	Colluvial
Ranker (Leptosol acidic, dystric)	Based on siliceous parent rock—origin of mineral component of soil
Rendzina (Leptosol calcaric)	Based on calcareous parent rock—origin of mineral component of soil
Chernozem	On loess, typical, calcareous
	On fluvial calcareous deposits
Vertisol	Based on parent rock—origin of mineral component
Class III. Cambic soils A–(B)$_o$–C or A–(B)$_o$–C–R or A–(B)$_o$–C–R soil horizons sequence	
Cambisol eutric	Based on parent rock
Cambisol dystric	Typical
	Podzolic (Brunipodzol)
	Pseudogleyic
Class IV. Polygenetic cambisols of karst A–(B)–R soil horizon sequence	
Kalkocambisol	Typical
	Luvic
Red soil (terra rossa)	Typical
	Luvic
Class V. Eluvial illuvial soils A–E–B–C/R and A–E–IIB–R frame of horizons	
Luvisol	Based on parent rock—origin of mineral component of soil
Podzol	
Acrisol (two layers)	Typical
	Pseudogleyic
Class VI. Autochtonous terrestrial anthrosols P–C, P–(B)–C, P–B–C frame of horizons	
Rigosol	Based on original soil
Hortisol	
Soil of green or plastic houses	
Soil of terraces	

(continued)

Table 5.4 (continued)

Type	Subtype
Karstanthrosol	Manual or mechanized cleaning/removal of stone/rock
	Mechanized removal of stone—deposited on soil surface
	Mechanized removal of stone, cracking by rotavator on surface
	Grinding of stone in situ, mixed with soil and natural vegetation and deposited on surface
Class VII. Technogenic soils (Technosols) I–II–III–IV…etc. layers	
Deposol	Deposited soil material from excavation in building activities
	Mining and barren waste from mining
	Ash and dross after burning of coal
	Municipal waste
Soil on flotation sludge	Mud sedimented in equipment for cleaning of waste water
	Ash and dross of burning coal in thermoelectric plant
	Mud excavated by cleaning and deepening of lakes, harbors, etc.
	Waste of mineral fertilizer industry, such as phosphogypsum
	Waste in lime production
Atmospheric deposition	Ash and dust emitted from industrial chimneys
(B) Order of semiterrestrial soils	
Name and main characteristics of soil taxons	
Class I. Stagnation of rainwater: A–E_g–S–C	
Pseudogley	On terrace
	On slope
	On lowland
Class II. Anthrosols semiterrestrial P–S–C	
Rigosol	Based on subtype of pseudogley
Drained pseudogley	
(C) Order of hydromorphic soils	
Name and main characteristics of soil taxons	
Class I. Initial multilayer soils I–II–III–IV…. etc. -$G_{r\ or\ ox}$ (within 1 m)	
Fluvisol	Calcareous
	Non-calcareous
Class II. Semigleyic soils A–… C–G_r (75–100 cm)	
Semigley	Based on horizons in upper, terrestrial layer
Class III. Hypogleyic soils A G_{so} G_r soil horizon sequence	
Hypogley	Calcareous, clayey (vertic)
	Non-calcareous
Humogley	Calcareous, loamy
	Clay, vertic
Class IV. Amphygleyic soils A_a–G_r–G_{so}–G_r (to 100 cm)	
Epigley	Light (sandy) soil
Pseudogley–gley	Loam
Amphygley	Clay vertic
Class V. Peat soils of T–G_r soil horizon sequence	

(continued)

Table 5.4 (continued)

Type		Subtype
Histosol		Shallow T horizon 30–50 cm
		Moderately deep T horizon 50–100 cm
		Deep T horizon 100–150 cm
		Very deep T horizon >150 cm
Class VI. Anthrosol hydromorphic P-…-$G_{r/so}$		
Rigosol		Based on original of hydromorphic soil
Drained soil		
(D) Order of halomorphic soils		
Name and main characteristics of soil taxons		
Class	Type	Subtype
Salty soils	Solontschak	
II. Alkalized soils	Solonetz	
(E) Order of subaquatic soils		
Name and main characteristics of soil taxons		
Underwater soils, soil on the bottom of a water body	Dy	
	Gyttja	
	Sapropel	

and water saturation occur, dominated by anaerobic conditions and reduction processes and periods of low-water-dominated aerobiotic conditions and oxidation processes. The cause of their occurrence is a compacted horizon, impermeable to rainfall water, where water after rainfall stagnates. In some older soil classifications these soils were sometimes classified in terrestrial but sometimes in hydromorphic soils.

The current classification introduces the new order of semiterrestrial soils.

The importance of hydromorphic soils is illustrated by the fact that along with semiterrestrial soils they occupy 30 % of the Croatian land area. The classification of hydromorphic soils in particular is a very sensitive issue. Specifically, if the criteria for taxonomy assess the origin of hydromorphism and the duration of full saturation of soil or some horizons, their place in the taxonomy could suggest the direction and techniques of hydromelioration of these soils. Namely, the type and intensity of hydromorphism direct investment into expensive drainage systems, necessary for stable agricultural production. It should be noted that endomorphology is exclusively used for the assessment of the origin and intensity of hydromorphism, which means the features of gley horizons, i.e., the type and extent or depth of occurrence of a gley horizon in the soil. The problem is the fact that endomorphology changes slowly and long after a change in the water regime and can be misleading as a criterion for evaluation of conditions in the soil. A far more reliable indicator of hydromorphism is the oxidoreduction potential of the soil in each horizon. More research should be undertaken in this direction. Another problem is that some soils, especially soils on limestone and dolomite, contain little phosphate and iron, the "material" that gives the bluish color to the gley horizon in which reduction predominantly occurs. Continuing Table 5.4, we find the taxonomy of the order of hydromorphic soils.

Salty soils in Croatia have a negligible extent and hence negligible importance, although it is realistic to expect that aridification of climate, which from the results of our studies we are witnessing today (Bašić and Franić 2003), may intensify the processes of salinization and alkalization.

Areas under water, including freshwater ponds for fish breeding and drinking water reservoirs, have a large area in Croatia. The trend of increasing demand for fish opens up the need for expansion of water reservoirs, although the water regime of the river that supplies such reservoirs brings into question the survival of existing ponds. Some are sure to leave shallow ponds dry. Silt is deposited on the bottom of such water bodies, thus resulting in a subaquatic soil.

References

Bašić F (1976) Pedologija, Viša poljoprivredna škola Križevci, udžbenik za studente, str. 231, Križevci

Bašić F (1982) Pedologija, Viša poljoprivredna škola Križevci, udžbenik za studente II. preuređeno i dopunjeno izdanje, str. 181, Križevci

Bašić F, Franić R (2003) Croatian agriculture at the crossroads, nature and man in Croatian Agriculture. Croatian food and food processing industry, PRO—TIM, Zagreb, p 82

Bašić F, Bašić I (2007) Navodnjavanjem tala suzbiti opustinjavanje Iloka!, Gazophylacium, God. XII br. 1–2 str. 73–93, Zagreb

Bogunović M, Vidaček Ž, Husnjak S (2001) Soil classification for the needs of regional planning in Croatia. Contribitions to the International Symposium "Soil Classification 2001". European Soil Bureau - Research report No. 7, p 63–67

Gračanin M (1951) Pedologija – III dio – Sistematika tala, udžbenik Sveučilišta u Zagrebu, Školska knjiga, str. 298, Zagreb

Husnjak S, Rubinić V, Vrbek B, Špoljar A (2010) Važnost, načela i pravila svjetske referentne osnovice za tlo (WRB) s primjerima korištenja u hrvatskoj, Agronomski glasnik. 63, 5–6; str. 347-365

Husnjak S (2012) Sistematika tala Hrvatske. Zavod za pedologiju, Sveučilište u Zagrebu, Agronomski fakultet, Zagreb

Kišpatić M (1877) Zemljoznanstvo obzirom na šumarstvo i gospodarstvo, troškom Kraljevske hrvatske zemaljske vlade, Zagreb

Kovačević P, Kalinić M, Pavlić V, Bogunović M (1967) Detaljna klasifikacija tala I izrada pedološke karte hrvatske Posavine, Kongres JDPZ, str 247–259, Zadar

Kovačević P, Kalinić M, Pavlić V, Bogunović M (1972) Tla Gornje Posavine. Institut za pedologiju i tehnologiju tla Poljoprivrednog fakulteta Sveučilišta u Zagrebu, Zagreb

Nestroy O et al (2000) Systematische Gliederung der Böden Österreichs (Österreichische Bodensystematik 2000), Heft 60, Wien

Neugebauer V, Filipovski G, Ćirić, M, Živković M (1963) Klasifikacija zemljišta Jugoslavije, VIII Kongres JDPZ, Ohrid

Racz Z (1999) Aktualna pitanja i problemi pedoloških istraživanja u svijetu i kod nas, Poljoprivredna znanstvena smotra, Vol. 64, No.3, str. 231–241Opatija

Resulović H, Čustović H, Čengić I (2008) Sistematika tla/zemljišta, Univerzitet u Sarajevu, str. 234

Šandor F (1911) Sammlung von Bodenkarten Kroatiens und Slavoniens und bilder uber die Kultivierung dieser Bodenkarten. Verhandlungen der II. Internat. Agrogeologenkonferenz, Stockholm

Šandor F (1914) Tloznanstvo I (Općena nauka o tlu), u izdanju Kluba hrvatskih šumarskih akademičara, str 164, Zagreb

Škorić A, Filipovski G, Ćirić M (1973) Klasifikacija tala Jugoslavije. Zavod za pedologiju Fakulteta poljoprivrednih znanosti Sveučilišta u Zagrebu

Škorić A, Filipovski G, Ćirić M (1985) Klasifikacija zemljišta Jugoslavije, ANU BiH, knjiga 13, Sarajevo

Špoljar A (1999) FAO klasifikacija s bazom podataka za pedološku kartu Republike Hrvatske sitnog mjerila Agronomski fakultet Sveučilišta u Zagrebu, magistarski rad, str.121 Zagreb

Špoljar A, Kušec V, Kamenjak D, Kvaternjak I, Peremin-Volf T (2006) Promjene pedoloških značajki đurđevačkih pijesaka uzrokovane revitalizacijom, Agronomski glasnik br. 3, str. 181–198

Vrbek B, Husnjak S, Pilaš I (2008) Classification of lowland Oak Forest Soils in Croatia. The International Conference & Field Workshops on Soil Classification, Santiago de Chile

WRB (2006) World reference base for soil resources World soil resources reports 103, Food and Agriculture Organization of The United Nations, p. 128, Rome

6 Pedosphere of Croatia

Using data from the General Soil Map (GSM) of Croatia at 1:50,000 and regional monographs (Kovačević et al. 1972; Škorić et al. 1977, 1987, 2003) data of Bogunovic et al. (1996) from the Department of Soil Science, University of Zagreb, Faculty of Agriculture, which is entrusted with keeping records of the GSM Croatia, created a digitized soil map of Croatia at a scale of 1:300,000. To create the different maps AutoCAD, Arc View, and ArcInfo software were used. The data from these maps were used for the development of the regionalization of Croatian agriculture and for various works on the subject (Bašić et al. 2003, 2007), which are used in this book (Fig. 6.1).

The map has 65 individual map units, each of which is composed of further pedosystematic units, demonstrating the great diversity of the structure of the Croatian pedosphere, described as a consequence of very complex factors of pedogenesis. However, the structure is dominated by several types of soil. The most common soils are shown in Fig. 6.2, including 12 pedosystematic units whose distribution is greater than 2 % of territory and eight with a presence of less than 2 %. Of course, the importance of soil does not depend (only) on its distribution. The most common soil is luvisol, followed by pseudogley and gley soils.

This means that the majority of Croatia is covered by soils at the climax stage of eluvial–illuvial soils, i.e., the last stage of the evolutionary series on coarse parent material.

6.1 Order of Automorphic Soils

The automorphic soil order is predominant with an area of 3,153,432 ha or 56.63 % of the Croatian land area, showing that it absolutely dominates in terms of surface representation in the pedosphere of Croatia. The genesis of these soils occurs under the influence of rainwater only, whose distribution shows an increase from east to west in the continental areas, in which lie the highest mountains in the region, and from north to south in the Adriatic agricultural regions. In recent years climate change has led to recurrent droughts and the initial processes of desertification in the eastern Pannonian subregion (P-1), and a corresponding shrinkage in the high rainfall intensity. Proper growth of precipitation and a temperature drop is reflected by the structure of the pedosphere, in which, from east to west on a substrate which originally was the same, loess of eolian origin, a climate zonality of soils, is evident, which was observed in the Croatian pedological literature and described by Janeković (1963). The following is a description of the soil types in the evolutionary order of the series, which are listed in Table 5.4.

6.1.1 Class I: Initial Automorphic Soils of (A)–R or A–C Sequence of Horizons in Soil Profile

Soil in this class is at an initial stage of genesis. At this stage it is affected by water erosion and/or wind, constantly removing soil, or covering formed layers with a new coat of soil material. Their main characteristic is that they are underdeveloped—with an initial (A) horizon, which can grow on hard rock, detritus of hard rock (regolith), or on a loose parent substrate.

6.1.1.1 Lithosol

The name of the soil type. Its name comes from the word lithos—rock, because stone residue of the parent substrate prevails on the surface, which of course affects all characteristics of the soil. This soil type is described primarily in numerous works of Croatian forest Soil scientists; (Martinović 2000; Martinović et al. 1990; Mayer et al. 1981; Pernar 1997; Miloš 1983, 1985).

The WRB name. Leptosol (Lithic, Calcaric, Dystric, Eutric, Haplic)

Genesis and sequence of horizons. Lithosol is called substrate-generated soil—the "youngest" member of the evolutionary series of soils on hard rocks of the profile (A)–R or (A)–C–R. The initial (A) horizon is interrupted by

F. Bašić, *The Soils of Croatia*, World Soils Book Series,
DOI: 10.1007/978-94-007-5815-5_6, © Springer Science+Business Media Dordrecht 2013

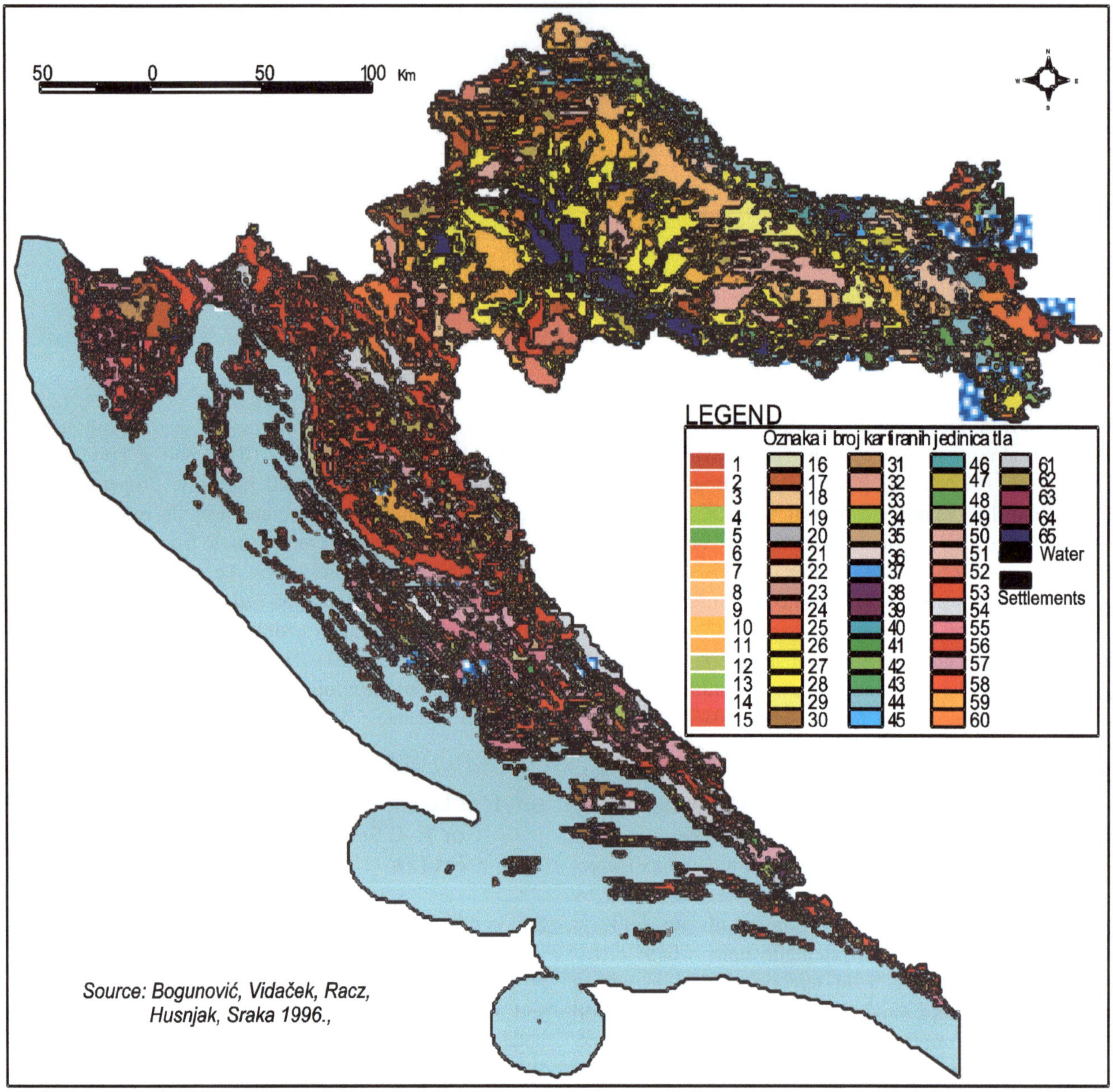

Fig. 6.1 Soil map of Croatia (1:300,000)

fragments of angular rocks or stones on the surface of the soil. The most common substrate is limestone and dolomite from different periods of the Mesozoic. In the initial stage of development it is affected by water and/or wind erosion, and this process is enhanced by excessive livestock grazing, which suppresses the already meager vegetation, bushes, or underbrush between rocks.

Position in an evolutionary series. This soil on stable topographic positions gradually evolves in the direction of kalkomelanosol or cambisol—*dystric* if the parent substrate rocks are acidic; *eutric*, if the rocks are rich in bases or carbonates, and on limestone and dolomite into kalkocambisols.

Distribution. It is widespread at around 32,703 ha or 0.6 % of the land surface of terrestrial Croatia. The spatial distribution is shown in Fig. 6.3.

In the Pannonian region there are practically none of these soils. A small amount occurs in the submountainous and mountainous regions, with the highest occurrence in the Adriatic agricultural region due to erosion by wind and water. This soil type can be prevalent on the islands, especially those exposed to strong wind, like Cres, Krk, Pag, Vis and Dugi otok, and in the hinterland of Zadar (Bašić and Šimunić 1985). Minor occurrences occur as inclusions across the whole Adriatic region.

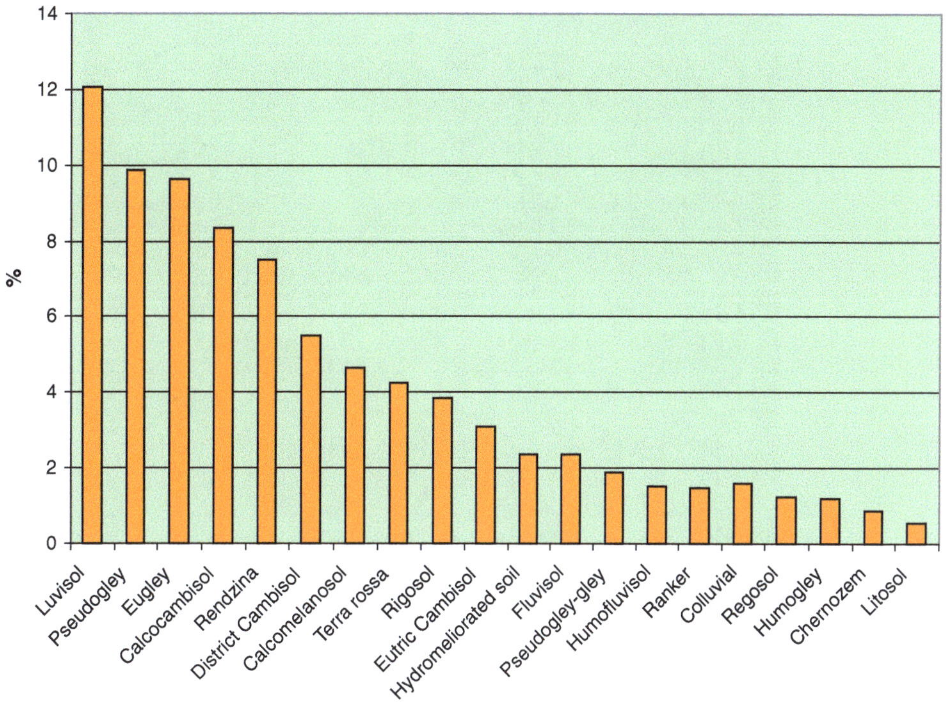

Fig. 6.2 Distribution of most widespread pedosystematic units in the Croatian pedosphere. *Source* Husnjak et al. (2005)

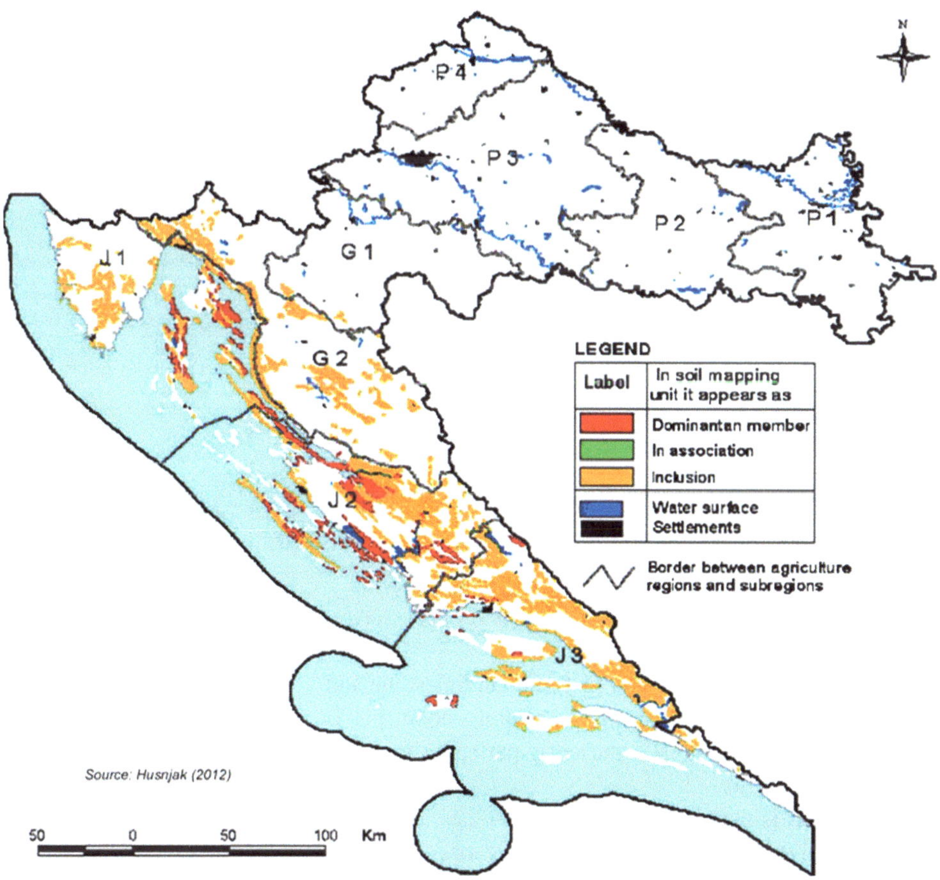

Fig. 6.3 Distribution of lithosol in the pedosphere of Croatia

Physical and chemical properties. It is a shallow, dry, and warm soil with a little fine soil material rich in humus. All properties depend on the characteristics of the substrate, the way it was fragmented, and the depth of the regolith.

Photo 6.1 The roots of forest vegetation in the mountainous region are found on the surface in evolving lithosols, but they also penetrate into cracks and "pockets" of deeper layers

Photo 6.2 Penetration of roots into deep layers is key to the survival of orchards. The photo shows an almond plantation in a field near Knin

This applies also to its chemical features. On limestone and dolomite the chemistry is practically neutral—about pH 7. Naturally, it is poor in all nutrients, particularly plant-available phosphorus.

Land use. It is mainly under vegetation—i.e., as pasture of low nutritive value on rocky, maquis, garrigue vegetation, or bare rocks. Because of its rich biodiversity it is used as pasture for sheep and/or goats. The livestock provide excellent meat and milk from the sheep pastures, and the dairy products, especially cheese, are superb delicacies. In the summer the above-ground part of the vegetation dries out, while in autumn and especially in spring it becomes green again. Forest vegetation can be found on parts of these soils. However, the roots of the vegetation penetrate into deeper layers, in the cuts and "pockets" of soil that receive water and nutrients (Photo 6.1).

Softer limestone (Eocene) creates a much deeper layer of fragments in which plants that require the more modest

Photo 6.3 Mountain pastures on lithosols in the Mountainous region (G-2), Kupinovo near Srb (*left*) and the central karst region in the hinterland of Zadar (*right*)

xerothermic conditions of such habitats, such as grapes and almonds, can survive and produce fruits of high quality (Photos 6.2, 6.3).

Guidelines for management. The management of these soils should include protection from water erosion and wind, inflow of new material and overgrazing, thus facilitating their evolution towards humus-accumulative soils—kalkomelanosol or rendzina.

6.1.1.2 Regosol

The name of the soil type. Its name comes from the word *rhegos*—blanket.

Genesis and sequence of horizons. It is an undeveloped soil with an (A)–C horizon sequence when formed on loose unconsolidated parent substrate, and (A)–C–R on marl limestone, marl, and flysch. The soil represents the first stage of evolution of soil on loose substrates affected by erosion, which prevents normal evolution and pedogenesis, and thus keeps the soil in the initial stages of development and evolution. Horizon (A) is shallow, abrupt and has a slightly darker color than the substrate, due to the modest content of humus, whose quality depends on the quality and composition of the parent material.

The WRB name. Leptosol

Position in an evolutionary series. Regosol is an "initial" phase. Namely, the discontinuous (A) horizon begins soil genesis and evolution. The direction is towards a humus-accumulative soil type A–C or R sequence of horizons.

Land use. As a consequence of anthropogenic impact through processing, bare soils are exposed to stronger rains, especially on slopes, which is conducive to water erosion. As these conditions occur more favorably in broad row crops, such as corn, its large representation in the Croatian fields raises the question of effective protection of the soil.

Distribution. As the distribution of loose substrates occurs in all regions, this soil is seen across the country with an area of 70,698 ha or 1.3 % of the Croatian land surface. Its spatial distribution can be seen in Fig. 6.4.

All properties of regosol depend on the characteristics of the parent substrate. If the parent substrate has favorable properties, the soil is usable for plant growing.

Regosol on loess occurs in the eastern Pannonian subregion (P-1) as the prevalent soil type in some mapping units. With a calcareous and favorable substrate that receives and retains enough rainwater, regosol is a soil of value. In some areas, especially the hilly, undulating and inclined positions characteristic of eolian geomorphology, such as the Vukovar loess terrace, Baranja hill slopes, and Fruska hill in Ilok, regosol is the dominant unit in the structure of the pedosphere. Water erosion has a decisive influence on its formation. The process begins on bare soil during heavy rains, when it first forms small rills which gradually increase, taking away the soil material of the entire plowed arable layer, and calcareous loess breaks the surface with its characteristic yellow color (Photo 6.4).

All the features of this soil type depend on the characteristics of the substrate. The initial plowing next to the

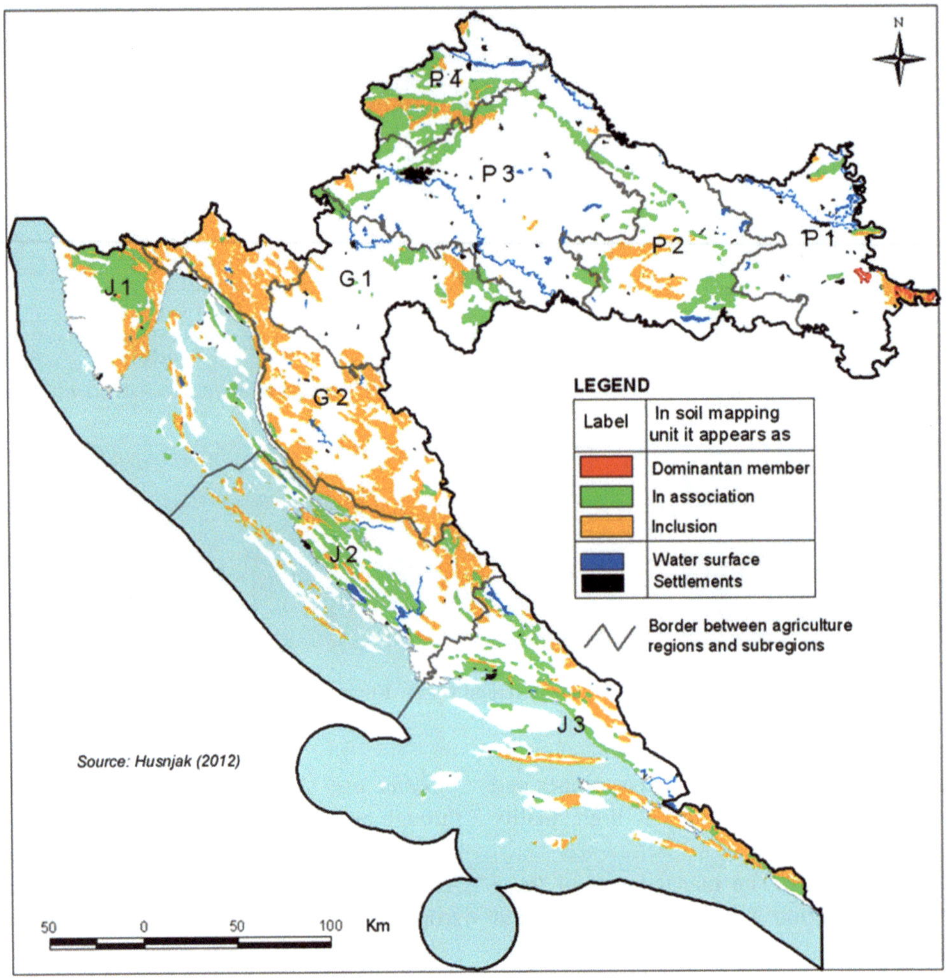

Fig. 6.4 Distribution of regosol in the pedosphere of Croatia

Photo 6.4 The initial stage of erosion of arable land in the locality Ilok—Principovac (P-1 subregion). Torrential flow of rainwater removes the plowed horizon, and loess breaks the surface—regosol occurs

surface of the substrate ejects loess, which, when mixed with plant residues, causes humification of plowed plant residue and the formation of an (A) horizon. This creates the arable land mosaic in the eastern Slavonian subregion,

Photo 6.5 Tillage of inter-row space in a permanent plantation with rows oriented in the direction of the slope near Novi Marof (P-4) favors water erosion of soils on marl, creating and maintaining regosol on marl (*left*). Water erosion "creates" the beautiful landscape of Under Kalnik with marl as the dominant parent material and regosol as the dominant soil type (*right*)

as described in the previous chapter, where in Photo 4.29 we see the light yellow color of regosol on loess, and eutric cambisol, which has the brown and dark color of a humus-accumulative horizon. However, after the spring the crop is green and covers the surface; we lose all traces in the "green sea" of crops. Regosol occurs in association with or as inclusions in fertile soils like chernozem or cambisol eutric on loess, which are very fertile soils with high yields of all crops.

Regosol on marl is widespread on the slopes of the mountains of the Middle Slavonian hills (Dilj, Papuk, and Krndija), slightly less so on the slopes of Moslavačka hill, and it is significant in the western and north-western Pannonian subregion (P-3 and P-4). We should mention Croatian Zagorje, Žumberak, and Plešivica in association with other soils, or, as an inclusion, topographic positions exposed to erosion, mostly where there is permanent row cropping with the rows oriented down the slope, without overgrowth of grass between the rows in these plantations. In addition, it also occurs in G-1 and the subregions across the Adriatic region, where marls occur. In breaks in vegetation one observes light-colored shale on the surface in the landscape.

Taking into account that pesticides are used as biocide chemicals on these soils, part of which ends up in the soil and is removed with the soil through erosion to streams, these soils may damage the water resource of an area (Photo 6.5).

As a parent substrate, marl has favorable properties. The characteristics of regosol on marl depend on the texture of soil (marl) and exposition of the slope.

Regosol on marl is basically a shallow, dry, usually light soil, permeable to water, alkaline, poor in humus and

Photo 6.6 Regosol on flysch at Buzet in Istria (J-1) with a relatively deep layer on sandstone, which prevents penetration of the roots into deeper layers

nutrients, particularly plant-available phosphorus. If "impurities" are present in the marl in the form of fragments of limestone the regosol could be stony (Photo 6.6).

Regosol on flysch is widespread throughout the Adriatic region and islands. In Istria it occurs all over the territory on surfaces regardless of their use in agriculture or forestry, because forest cover does not provide sufficient protection against water erosion.

These are loose, dry, and warm soils, with a relatively high content of clay minerals, which means a heavy texture, and permeability to water, which depends on the characteristics of flysch. As a rule, the regosol is a stony soil, because of the fragments of sandstone. The fertility of this soil depends on the depth of the layer of sandstone.

We can find regosol to a lesser extent on other parent materials.

Regosol on Pleistocene loam is the most widespread soil type in the P-2, P-3, and P-4 subregions and on the slopes of Bilogora, the Middle-Slavonian hills, and Hrvatsko Zagorje.

Regosol on sand we find on eolian sands, on the slopes of Bilogora near Đurđevac, all under the influence of mainly eolian erosion. In the Adriatic region, on the mainland in the hinterland of Zadar (flat districts) and on the islands, especially on the so-called Cres saccharoides dolomite we see the formation of regosol on dolomite sand, giving a dry soil covered with vegetation, bushes, or underbrush.

Position in an evolutionary series. It is clear that regosol occurs in all agricultural regions and subregions, in topographical areas conducive to soil erosion. If appropriate interventions suppress the erosion of soil, the evolution moves in the direction of the formation of a humus-accumulative molic–A_{mo} horizon, i.e., shallow rendzina on marl, dolomite, flysch, sands, etc. On loess, the evolution is directed towards chernozem; on Pleistocene loam that was leached, washed out, and that has an acid chemistry, the evolution leads towards eutric cambisol.

Land use. The soil is used as a more or less suitable substrate for perennials—vineyards and orchards. As regards production value and other characteristics, regosol is a very heterogeneous soil. The production value is good up high, which mostly depends on the depth of the substrate or the possibilities of the root to penetrate deep into the soil mass. It is significantly affected by climate conditions: the amount and distribution of rainfall. The most valuable is regosol on loess, and the least productive is that on sand and unconsolidated fragments of silicate rocks. Generally, their natural nutrient content is insufficient; there is little biological accumulation of nutrients due to the low content of humus. A positive characteristic is the great physiological depth of soil and root penetration, making deep rooting in the active profile of the substrate possible. Regosol on loess is a good substrate for very valuable intensive viticulture (Ilok vineyards and Baranja) and fruit growing. Regosol on marl and flysch are also good substrates for viticulture. Našice, Jaska-Plješivica, Kalnik vineyards, Međimurje and Zagorje are mostly situated on these soils. Part of the vineyards of Istria, the Dalmatian coast, and the islands are also located on regosol. A negative and very unpleasant feature of regosol on marl and flysch is its tendency to landslides. This process is promoted by the work of heavy equipment, especially vibrating machines that disturb the deeper layers, resulting in cracks that receive larger quantities of water after heavy rains, increasing the load-slip in the direction of inclination of the slope.

Guidelines for management. First one should combat erosion, thereby facilitating the evolution of the soil to the next developmental stage, and therefore evolution to a humus-accumulative soil. The prerequisite for this is stability of the land to protect the soil from erosion and landslides. The first and easily feasible measure with this goal in mind is contour management—tillage and rows of permanent plantations in the direction of contours. This method of soil tillage allows the application of modern machinery to gradually form terraces that can without consequences accumulate the water from rains of high intensity. Influenced by climate change, such rains are today a common event. Of course, contour tillage and management opens the question of rows that are in shadow, not exposed to sunlight. If this overshadowing is decisive for the quality of fruits and/or grapes, an alternative is planting between rows of appropriate plants or grass bands at set distances. Our calculations showed that the soil-loss tolerance (T) for the regosol on loess in Ilok is only 4 t/ha/year. On this basis, we calculated that every inclination of more than 3 % for any length of slopes greater than 10 m requires "green cover" in the inter-row spaces of vineyards (Bašić et al. 1999, 2005, 2007, 2009).

For different soil types, of very high importance is the maximal length of plot on which one should create barriers to stop eroded soil. According to our calculations for efficient soil conservation in the P-1 subregion, contour tillage and contour rows of vines will be a very effective measure.

Photo 6.7 Fero-chlorosis of grapevine on regosol on marl—Kalnik (P-4). First, it is manifest in young leaves and shoots, and then it affects the older leaves

Typically, these soils have a low supply of plant nutrients, so fertilization and humification combined with deep tillage is the main intervention in management of this soil. It is advisable to use legume species as green manure crops.

On the calcareous (calcium-rich) substrates, particularly in regosol on marl, there is a frequent occurrence of jaundice caused by ferro-chlorosis—iron deficiency chlorosis. To establish vineyards or orchards on such substrates very deep plowing of the soil is not advisable, due to the ejection to the surface of the substrate which is rich in $CaCO_3$. Regosols on silicate substrates are very valuable forest soils (Photo 6.7).

6.1.1.3 Arenosol

The name of the soil type. The origin of the name of this soil type is from the Latin *arena*, sand, and means sandy soil. In the Croatian language people call these soils "living sand," because they are in a state of permanent movement.

Genesis and sequence of horizons in soil profile. Arenosol is an undeveloped soil of (A)–C horizon sequence formed on eolian sands permanently being moved by wind. It occurs at several localities along the river Drava and at some localities in the Mountainous region.

The WRB name. Arenosol

Position in an evolutionary series. Arenosol is the initial–early stage of soil genesis, whose development is affected by water erosion. Development is possible only after procedures to calm the sand and application of windbreaks. Then evolution continues to a humus-accumulative soil—leptosol with an A–C soil horizon sequence.

Distribution. It occurs over a negligibly small area of 667 ha or 0.012 % of Croatian territory (Fig. 6.5).

The largest area is located near Đurđevac and a protected area—the botanical reserve called Đurđevac sands, known under the name "Croatian Sahara." Other sites of arenosols are around Virovitica and Slatina. Prior to the soils being afforested and conserved in the first half of the twentieth century, during dry years after a long period of drought the wind was able to eliminate the vegetation cover, which protects the sand from movement.

Physical and chemical properties. The characteristics of the soil naturally depend on the characteristics of the sand.

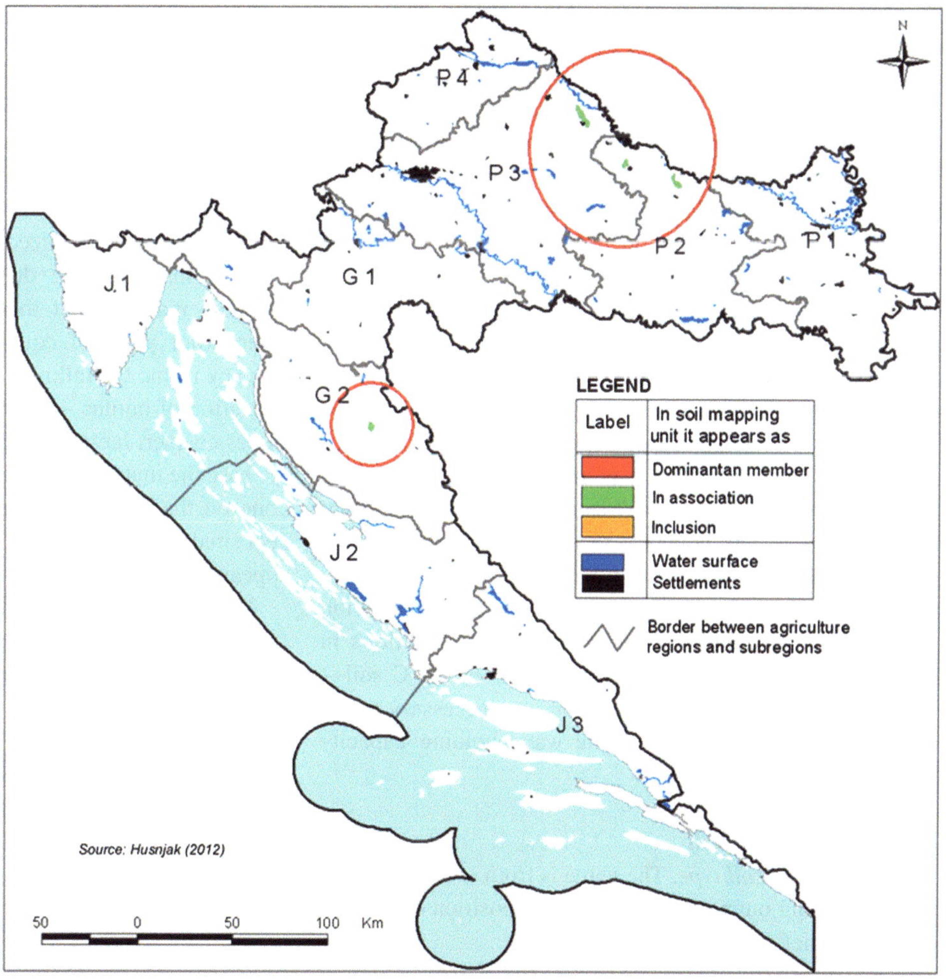

Fig. 6.5 Distribution of arenosol in the pedosphere of Croatia

Photo 6.8 Arenosol in the botanical reserve Đurđevac sands—(*left*) *Sarothamnus scoparius* in bloom (*right*)

Siliceous sands are acidic and of low fertility, while siliceous carbonate is much more favorable. As soon as winds calm, in the mixed surface layer different sands can be found. Because of the high permeability of the sand, bases are flushed/leached and the acidification process is quite intense. Sands at the soil surface can be quartz with over 95 % quartz, siliceous with below 95 % quartz, or carbonate-siliceous with less than 95 % quartz and some $CaCO_3$ content (Photo 6.8).

Land use. These soils are dry and warm habitats of special psammophytic vegetation. They are used as arable land, and arenosol occurs only sporadically on "tops" of eolian dunes—the hills where in dry years the crop is poorer because of water deficit. Most often *Robinia pseudoaccacia* occupies this position, which is also an important species for the conservation of "calm" sand. Arenosol is also used as acacia forests and botanical reserve fields.

Guidelines for management. The soil needs protection from wind erosion, and thus needs planting with vegetation species that can withstand the hot and dry conditions in these soils. This should allow evolution of the A–C soil—leptosol. On arable land with this soil, it is necessary to use organic fertilizers to increase the water-holding capacity (Špoljar et al. 2006).

6.1.1.4 Colluvial Soil

The name of the soil type. The name is from *colluvic*, which means washing out with transport and deposition on a slope or stable relief form.

Genesis and sequence of horizons. Colluvium is the undeveloped phase of soil genesis which is caused by recent sedimentation of soil material mixed with fragments of rock moved from higher relief positions to a lower one by flood or surface water runoff. They form an (A) or $(A)_p$ horizon on arable land. The eroded material accumulates in the lower position of slopes, on the bottom or in the valleys, where it can be influenced by underground water. Certain layering may be seen due to the unequal deposition of colluvial material, but there may be no difference in the layers after years of equal, homogeneous material flows. The key is that the influx of new material does not allow the formation of humus—the A horizon. On the upper parts of slopes coarser, larger, skeletal material is deposited, and on the base the material is finer. The division into lower units is done on the basis of the properties of the material that is deposited—eutric, dystric, skeletal, calcareous, acidic, appears gleyey from groundwater, salinized water, etc.

The WRB name. Colluvial soil

Position in an evolutionary series. The evolution of these soils occurs in the direction of the formation of an A–C soil horizon sequence, which means a ranker or rendzina (Fig. 6.6).

Distribution. This soil type we find throughout the state in the area of over 91,938 ha or 1.65 % of the Croatian land area. It rarely occurs as the dominant unit within cartographic units. Larger areas are at the foot of the Middle Slavonian hills in the Pannonian region and in the karst areas of the Adriatic region. It is far more widespread in

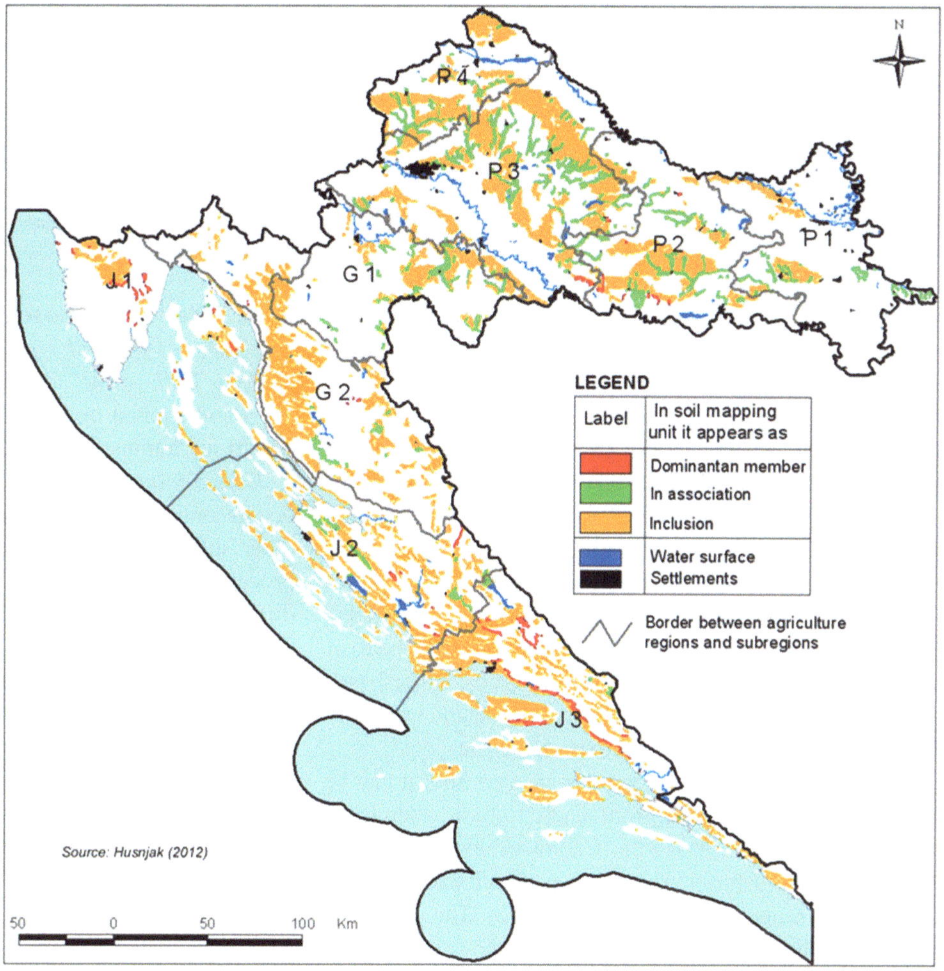

Fig. 6.6 Distribution of colluvial soils in the pedosphere of Croatia

Photo 6.9 *Left* Profile of colluvial soil on limestone karst, which is deposited and stabilized in negative relief forms (photo: Husnjak). *Right* A freshly plowed colluvial soil: Cres city on Cres island

association with a variety of hydromorphic soils in lowland areas as well as in the form of inclusions across the country (Photo 6.9).

Physical and chemical properties. All properties are very heterogeneous and depend firstly on the origin of the soil material. The texture of colluvial soil ranges from skeletal to sandy to clayey soils. As a rule, they are well drained and aerated, warm, and have a low water capacity. Therefore, in dry years, crops on this soil suffer from drought. In contrast, in the zone of pseudogley and luvisol in the Pannonian region, colluvial soil is a fertile soil, in which water is generally not lacking, even in dry years. But in this soil type, deposited on the bottom of slopes with arable land, heavy metals, nitrates, and residues of pesticides translocated from the topsoil of the arable land can accumulate. Because of contact with water courses and groundwater for drinking water, it is necessary to bear in mind the possibility of pollution. In the P-2 and P-3 subregions, with very intensive arable farming and the usual use of high doses of fertilizers and farming with very narrow crop rotation (maize–wheat) or with monoculture of maize as profitable crops, water-based soil erosion of the top tilled layer after sowing and treatment by pesticides is an ordinary and expected event.

However, the chemical properties are versatile; it can be calcareous, non-calcareous, from acidic to basic chemistry, with humus content of 1–3 %. Cation exchange capacity (CEC) and base saturation depend on the character of the material and range from low to high values. The supply of plant-available nutrients is fair.

Land use. In spite of heterogeneity this soil type in the Pannonian region is a very valuable arable soil, usable for growing all types of arable crop. Very good quality valuable forests occur on this soil in the mountainous region, while in karst areas arable land or meadows are found. Productive features depend on the physical and chemical properties, especially the texture, depth, and content of the skeleton.

Guidelines for management. For the repair and stability of these soils, the most important intervention is protection from inflow of new material.

6.1.2 Class II: Humus-Accumulative Soils of A–C, A–R or A–C–R Soil Horizons Sequence

This class consists of young soils and represents the next stage of evolution after the initial one. Their soil profile is (A)–C or R, representing accumulation of humus as a dominant process of genesis. The effect of this process is usually estimated based upon thickness and type of humus-accumulative A horizon. All the features of these soils, especially those important for growing plants or the growth of natural vegetation, are dependent on the properties of the parent substrate. For example, a C-type substrate allows deeper penetration of plant roots and has a bigger volume of soil for the rhizosphere, whereas hard rock, i.e., R-type parent substrate, limits rhizosphere expansion. Conversely, a calcareous and base-rich parent substrate enables organic matter transformation in the direction of a humus-accumulative molic type (A_{mo}) horizon, which enables very favorable granular structure and soil properties.

6.1.2.1 Kalkomelanosol

The name of the soil type. The name originates from the German *Kalk*—lime, and *melanic*—dark-colored, suggesting a soil formed on limestone and dolomite of dark color due to the humus horizon.

Genesis and sequence of horizons. Kalkomelanosol occurs only on limestone and dolomite. In relation to all other substrates, these rocks as the parent substrate in the process of pedogenesis have features that are significantly different from other substrates. Specifically, for genetically younger soils on other substrates the influence of the substrate on the properties of the soil is high. The influence of limestone is quite different. Limestone mineral soils are formed from the entire substrate, but only from the so-called insoluble residues. The pure limestones of the Triassic and Jurassic contain over 99 % $CaCO_3$, which is chemically consumed and washed from the soil, and there remains a silicate residue, just below 1 % insoluble, from which the mineral part of the soil is formed. The dominant process of genesis of kalkomelanosol is dissolution and leaching of calcium carbonate and magnesium, with the formation of soluble bicarbonate and the accumulation of humus in the form of calcium and magnesium humate.

This is a soil of A_{mo}–R, or in forest O–A_{mo}–R, soil profile horizon sequence, which occurs on hard limestone and dolomite only. The humus-accumulative A or O horizon lies directly on hard limestone or dolomite rock or on gravel of limestone, containing fragments of limestone or dolomite. Numerous researchers have described this soil under different names: humus-carbonate soil, mountain chernozem, meadow and mountain black soil, etc. It occurs on all limestone and dolomite of the mountain regions, especially in the Dinaric mountain system from Ćićarija in Istria to Konavle. It occurs in association with kalkocambisols and luvisol on limestone (Pernar 1997).

The WRB name. Leptosol

Position in an evolutionary series. Soils on pure limestone are very old; they are polygenetic formations. The dissolution of limestone is very slow and soil formation on this rock occurs over a very long period. For example, for the formation of a layer of insoluble residue of only 1 cm thickness a time period of 8,000 10,000 years is required. The genesis varies with the different bio-climatic conditions and soil characteristics; therefore, the limestone is often not correlated with the actual conditions of pedogenesis, especially climate. Dissolution of limestone is a process that occurs with varying intensity in all climatic conditions. This is the reason why the first stages of development of soils on limestone are similar under different climatic conditions. In all stages of development on limestone, the soils have good permeability to water; thus they are not affected by the stagnation of water leading to marked textural differentiation of soil profiles. In relation to the particular surface and the "underground" relief of the parent material (cracks, sinkholes, and other morphological features, described as karst phenomena in the previous chapter), vegetation cover and erosion intensity (wind and water) vary, and the soils take on a distinctive mosaic pattern over a small area, with variable soil depth. Evolution of soils on limestone is highly correlated with soil depth, as this provides greater depth of moisture retention, increased biological activity, and the processes of mineralization of humus, which is then accumulated in the initial stages of development. Therefore, the limestone is generally present in an evolutionary series of soils:

LITHOSOL (beginning of the accumulation of humus) → **KALKOMELANOSOL**; **organogenic** (humus accumulation with small mineral component – insoluble residue of limestone or dolomite) → **organo-mineral** (along with humus accumulation of mineral component – insoluble residue) → **cambic** (increased mineralization of humus and the accumulation of mineral component) → **KALKOCAMBISOL** (greater depth of mineral component, decreased humus content because of intensive mineralization) → **LUVISOL ON LIMESTONE** (along with decreased humus content differentiation of clay content in profile).

In the initial stage it is dominated by the accumulation of humus, and organogenic kalkomelanosol is formed, with a lower mineral component content in the soil. By increasing the mineral content this also increases the water-holding capacity and the capacity for microbiological activity. Besides the accumulation of insoluble mineral residue due to favorable conditions for microbial activity, at this stage there is more intensive mineralization of humus and its content gradually decreases. The same tendency of increased intensity occurs in the kalkomelanosol cambic stage, which already has an initial $(B)_{no}$ horizon. Further evolution leads towards the genesis of kalkocambisols on limestone and dolomite. If the parent limestone is of Cretaceous age and the insoluble residue has a red color, evolution is driven in the direction of the formation of red soil (terra rossa). Also, strong water or wind erosion processes can form deep colluvial deposits. Classification according to Variety includes organogenic and organo-mineral, but classification based on Form depends mostly on topographic position, giving very shallow to very deep forms with significant differences in fertility.

Because of the permanent good infiltration of rainwater and soil drainage, as a rule the evolutionary stage pseudogley is absent. Given the fact that the insoluble residue has a kaolin clay character (dissolution of limestone, unlike argilogenesis) all soils formed on limestone (except kalkomelanosol) are rich in clay (Table 6.1).

Of course, the subtype colluvial is very favorable, with deep form, both the organogenic and the organo-mineral variety.

Distribution. Kalkomelanosol occurs over a respectable area of 255,201 ha, which represents 4.6 % of Croatian territory.

The dominant unit is on the mountain tops and highlands or in the high and cold karst fields of Ćićarija and Čabar over the Gorski kotar, Dinara, Lička Plješivica, Kozjak, Mosor and the tops of Brač, Korčula, and Hvar and the mountain heights above Konavle. These are the positions at which the loss of humus due to mineralization is halted in summer by drought and in winter by low temperatures, and humus accumulation is a fundamental process of genesis of kalkomelanosol. In association, mostly with other members of the evolutionary series on limestone and dolomite, it occurs along the entire Adriatic region and G-2 subregion, except of course for the area of flysch or igneous-metamorphic rocks. The humus-accumulative horizon is thick and full of vegetation roots, which provides very efficient protection of the soil against wind erosion (Photo 6.10).

Physical and chemical properties. Kalkomelanosol is a shallow soil, organogenic and, because of the small amount of mineral component, a silty soil, almost like powder in a dry state when it is easily dispersed and removed by wind. It is a dry, porous, warm soil, very permeable to water. Usually it is covered by xerophytic natural vegetation of high density—underground by root and above ground by a green mass. The main properties are illustrated in Table 6.2.

Although formed in limestone, a special form of genesis results in non-calcareous soil, the chemistry of which ranges from slightly acid to slightly alkaline. A very important remark in relation to soil sample collection and preparation in the laboratory is that non-calcareous soil material may contain very small fragments of limestone or dolomite brought in by wind and collected along with the soil sample. The consequence is that the pH will be higher than it is in nature! Kalkomelanosol is an acid soil in the western zones

Table 6.1 Classification of kalkomelanosol

Variety	Form
Organogenic	<15 cm (very shallow)
	15–30 cm (shallow)
Organo-mineral	30–60 cm (medium deep)
	>60 cm (deep)

Fig. 6.7 Distribution of kalkomelanosol in the pedosphere of Croatia

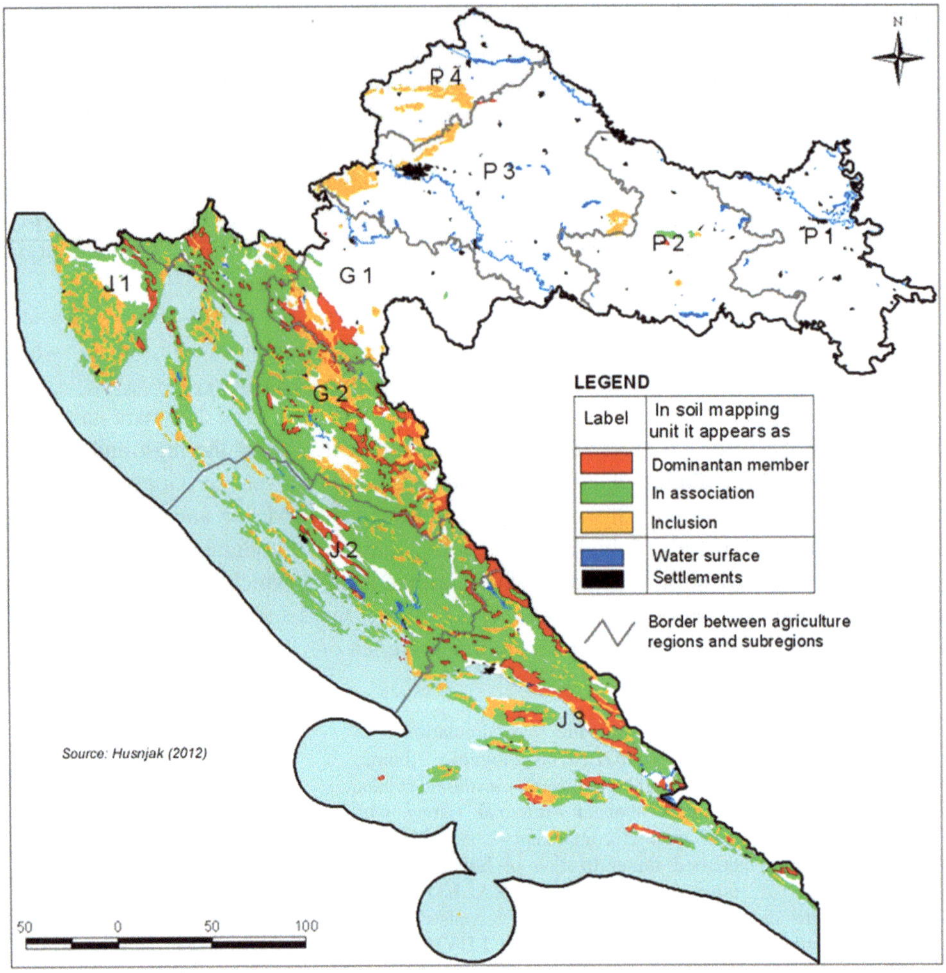

Photo 6.10 Kalkomelanosol on dolomite sand in the Mountainous area near Otočac (*left*) is an excellent substrate for the cultivation of potatoes and all root crops—vegetables, especially carrots of extremely high quality, in particular for baby food. When they are very shallow, these soils form lush mountain pastures at the top of Velebit (*right*)

and humid mountains. The humus content in the organogenic stage amounts to 60 %, and in the organo-mineral stage 10–25 %, because they have a high CEC (greater than 50 meq/100 g of soil), and are saturated with Ca^{++} and Mg^{++} ions. The kalkomelanosol on dolomite containing excess Mg^{++} is physiologically the driest soil. The supply of all nutrients is high in the soil. The total nitrogen content is high, but it is bound in the humus, and it is released slowly

Table 6.2 Physical and chemical properties of kalkomelanosol

Soil horizon	pH (nKCl)			mg P_2O_5/100 g of soil			mg K_2O/100 g of soil			Humus (%)			Clay content (%)		
	Average	Minimum	Maximum	Average	Minimum	Maximum	Average	Minimum	Maximum	Average	Minimum	Maximum	Average	Minimum	Maximum
A_{mo}	6.1	4.6	7.2	2.9	0.0	19.1	17.5	7.6	40.8	19.8	7.4	60.3	29.3	15.0	59.9

Source Bašić et al. (2003)

and in small quantities because of poor microbial activity, and extensively washed. Because of the good drainage kalkomelanosol is not subject to water erosion, but eolian erosion is dangerous in mountainous areas, especially on bare pastures where excessive grazing can exacerbate this process.

Land use. The largest black soil areas are under forest and pasture, the smallest are under fields, where it is excellent soil for potatoes, rye, and wheat. Forest vegetation develops roots in the surface soil layer, but due to the mosaic of soil cover and soil material, kalkocambisols and luvisol form in "pockets," cracks and crevices. Under such conditions, on the surfaces covered by kalkomelanosol are very valuable forest stands of beech, spruce, and fir.

Guidelines for management. This soil is valuable for arable farming if other conditions (e.g., length of vegetation period) are not limiting factors. If used in this way intensive fertilization with phosphorus and nitrogen is appropriate, and, for the growth of seed material of potatoes, with potassium fertilizers. The value of pastures on this soil type is very high, because livestock breeding based on pastures in our mountainous regions provides products of prime quality. Productive potential is not even remotely exploited. Fertilization by phosphorus and potassium in intensive pastoral or meadow management is a very profitable business.

6.1.2.2 Ranker

The name of the soil type. The name is derived from German: *rank*—steep slope, because the soil occurs just precisely in those topographic positions. It is traditionally described with this name in Croatian soil science.

Genesis and sequence of horizons. Ranker is a soil type of A–R or A–C–R soil profile horizon sequence, formed on noncalcareous acid silicate rocks. The main process is accumulation of umbric humus in an A_{um} humus-accumulative horizon. Depending on the features of the parent substrate we classify R as hard rock (lithic) or regolitic ranker, consisting of erosion-accumulated fragments of acid rock.

The WRB name. Leptosol acidic, dystric

Position in an evolutionary series. The genesis of ranker is led by processes of humus accumulation followed by intensive physical and chemical weathering of rocks, erosion, and leaching. The processes take place under cold climatic conditions with high rainfall quantity. The substrate on which the ranker is formed is very heterogeneous in composition, dependent on base content and CEC colloid complex base saturation. Base saturation results in different forms of humus, mainly acid and saturated by H^+, Al^{+++}, and Fe^{+++} ions. Because of the high permeability to water and intense leaching, the evolution of this soil leads to dystric cambisol. If this process takes place under very cold climatic conditions and high rainfall such as in Gorski Kotar on acid metamorphic rocks, sandstones and conglomerates quickly begin the process of podzolization, first releasing gray (the color of ash) sand visible only up close, and then forming an initial spodic E horizon. On steep topographic positions this may translocate and deposit, creating a very deep, fresh soil, an excellent substrate for the mountain forests of fir, beech, and sycamore maple.

Table 6.3 Classification of ranker

Variety	Form
For all subtypes • Typical • Cambic • Podzolized • Colluvial	For all varieties • Lithic • Regolitic • Deep (>60 cm)

Therefore, creation of a deeper A horizon leads to progression of base leaching processes and acidification, and starts the process of destruction of primary minerals and secondary mineral formation, resulting in the creation of the initial $(B)_o$ horizon, easily recognizable by its rusty-reddish color.

The direction of evolution can be seen from Table 6.3.

Distribution. Ranker occurs on siliceous acidic rocks of the Mountainous region (G-2 subregion) over an area of 86,944 ha, or 1.6 % of the Croatia land area (Fig. 6.8).

As the dominant soil type in cartographic units, it occurs on a mix of silicate gravel and sands in multilayer deposits of fluviative origin along the river Drava, in association with dystric cambisol on the metamorphics of Moslavacka gora, Psunj, Papuk, on the sandstones and metamorphics of Petrova and Babina gora, and in Gorski Kotar (Zalesina, Crni lug). As inclusions, it occurs in karst regions in areas underlaid by acid metamorphics—silica quartz sandstone and conglomerate. The soil type is cultivated on quartz gravels that have been transported fluvially. Otherwise, rankers are typical forest soils, still very rich in humus, which is slowly mineralizing; therefore, an organic—O horizon often appears (Photo 6.11).

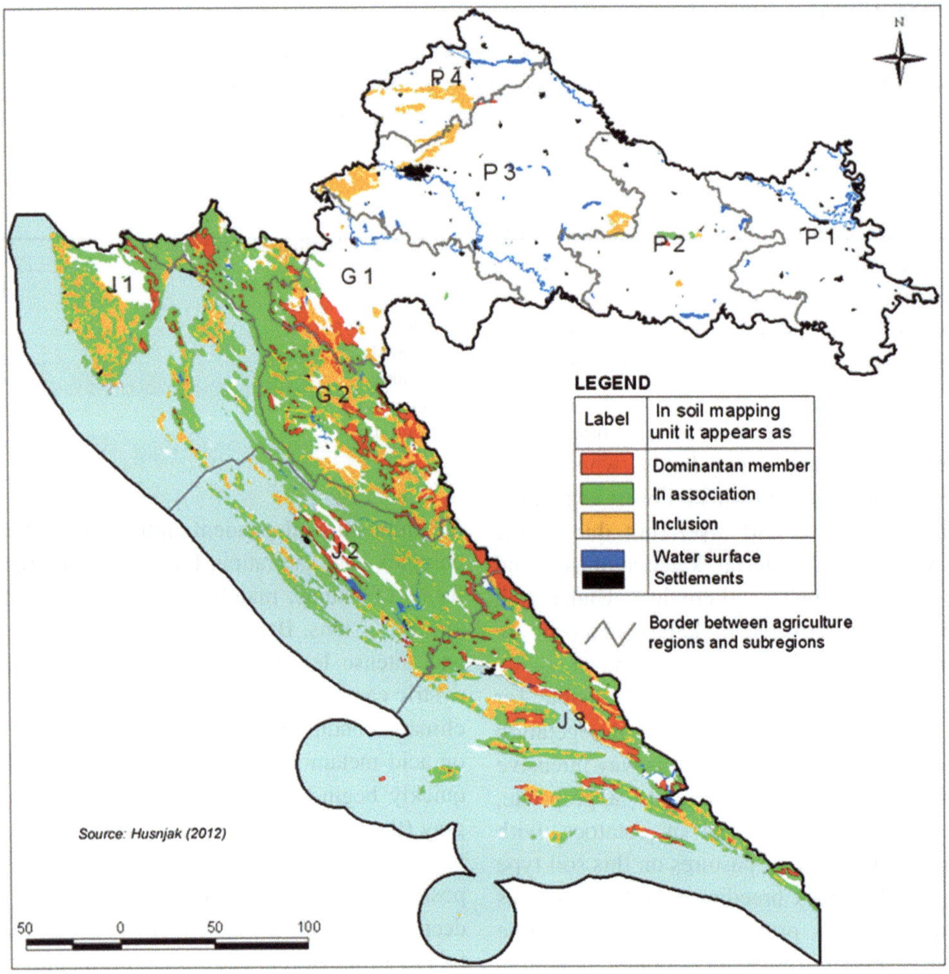

Fig. 6.8 Distribution of ranker in the pedosphere of Croatia

Photo 6.11 *Left* Ranker on quartz gravel and sand deposits of the Drava River (Petrijanec near Varaždin). *Right* Ranker on metamorphites of Kalnik (Vratno)

Physical and chemical properties. Ranker is a light, sandy, and skeletal soil with low clay content, well aerated and permeable to rainwater. Chemical properties depend on the properties of the rocks, but as a rule the chemistry ranges from strongly acid to neutral. It is very richly supplied with humus, has a high CEC, but low base saturation, and nutrient supply also ranges from poor to good.

Land use. One can observe that ranker is used mostly as forest soil in the mountainous region, but in plains and the Pannonian region ranker is arable land under cultivation.

Guidelines for management. If used as arable land, ranker first requires liming, with dolomite powder being the best material, because in the parent substrate there is no Ca^{++} and Mg^{++}. Liming is especially recommended on arable land used for growing the very famous "Varaždin cabbage" as well as fpr potato growing in Međimurje. Abundant fertilization with mineral fertilizers, with an emphasis on phosphorus, is recommended and very profitable.

6.1.2.3 Rendzina

The name of the soil type. The name originates from the Polish language, and was based on the characteristic of the sound when a hoe struck a stone: rndzzz! The literature recognizes this soil as the humus-carbonate soil. Earlier, in the older literature kalkomelanosol was described as Black soil on limestone.

Genesis and sequence of horizons. Rendzina is soil of A–C or A–AC–C soil horizon sequence, with a calcareous, molic A_{mo} horizon, formed on loose carbonate or silicate-carbonate substrates or on hard calcareous substrates, such as soft (Miocene, Eocene) limestone or marl limestone prone to fragmentation and dissolution. The substrate contains 10–50 % $CaCO_3$, which is seen on detailed observation. The main processes of pedogenesis are accumulation of mature, base-saturated humus and a colloidal complex in which Ca^{++} ions dominate, and formation of organo-mineral compounds. Morphologically there is a very clear remarkable molic A horizon of dark color because of the high content of humus, and a transitional AC horizon that is somewhat lighter in color. In the variety known as leached rendzina in the transitional AC horizon, accumulations of $CaCO_3$ or pseudo-spawn may occur. As a young soil in rendzina there are fragments of parent material, so all the features of the soil depend on the properties of the parent substrate.

The WRB name. Leptosol, calcaric

Position in an evolutionary series. Rendzina arises gradually from regosol on carbonate, easily disbursed substrates, and on topographically stable positions, after leaching of $CaCO_3$, and it evolves first into cambic rendzina (leptosol, cambic) and then cambisol eutric soil. In addition, since the rendzina is very permeable to water which washes out the carbonate present, the evolution of rendzina starts by leaching of $CaCO_3$ (decarbonatization). This process typically exceeds carbonate rendzina and there follows a variety of leached rendzina which opens the way for the gradual acidification and increased chemical decomposition of the primary minerals of the soil, creating first an initial cambic

Table 6.4 Classification of rendzina

Variety (based on $CaCO_3$ content)	Form (based on depth of soil)
Calcareous	>15 cm (very shallow)
Leached	15–30 cm (shallow)
Cambic	30–60 cm (medium deep)
	>60 cm deep

$(B)_o$ horizon—cambic rendzina, and in the next stage of soil evolution a proper cambic soil, thus:

Regosol → **Rendzina** (calcareous → leached → cambic) → **Cambisol eutric**

On soft limestone the last member of the evolutionary series is kalkocambisol. Classification of the lower units of the type is made as shown in Table 6.4.

Distribution. It occupies a respectable area of 420,184 ha of land or 7.55 % of Croatian territory, with the spatial distribution shown in Fig. 6.7.

As can be seen, rendzina occurs throughout the state, on flysch, marl, and calcareous and dolomite sand, but minimally in the P-1 subregion as inclusions, while in all other regions it appears as either the prevalent soil in a mapping unit, or in association with other soils, mostly regosol and cambisol eutric (Photo 6.12, Fig. 6.9).

Physical and chemical properties. We have calculated a range of important traits and their average value, and the results of the calculations are presented in Table 6.5.

Rendzina is loamy soil, but in lake sediments, like marl, clay content can be up to 50 %. Rendzina has generally favorable physical properties, it is stable, has an outstanding crumbly structure, a favorable ratio of pores, good water permeability, and favorable air–water relationships. As regards plant roots, the rhizosphere develops in the A and AC horizon and in some subtypes (e.g., on marl) where it enters the substrate in search of water. Rendzina is calcareous throughout its depth, except for the leached variety and cambic rendzina. The content of carbonate increases with depth. It has a neutral chemistry, and the leached and cambic varieties may be slightly acidic, which can reduce the saturation of the CEC. It contains 5–20 % humus in rangelands, and on machined surfaces this is generally much lower, because the processing intensifies mineralization. The supply of basic nutrients depends on the substrate, but usually lacks phosphorus.

Land use. Natural vegetation is xerothermic woods or rich pastures. Rendzina is a valuable agricultural land, much better than regosol on siliceous carbonate substrates, since it is deeper, richer in nutrients, biologically very active, and has much better air–water relationships. It is used as arable land for growing vegetables, grapes, and all kinds of fruit (Photo 6.13).

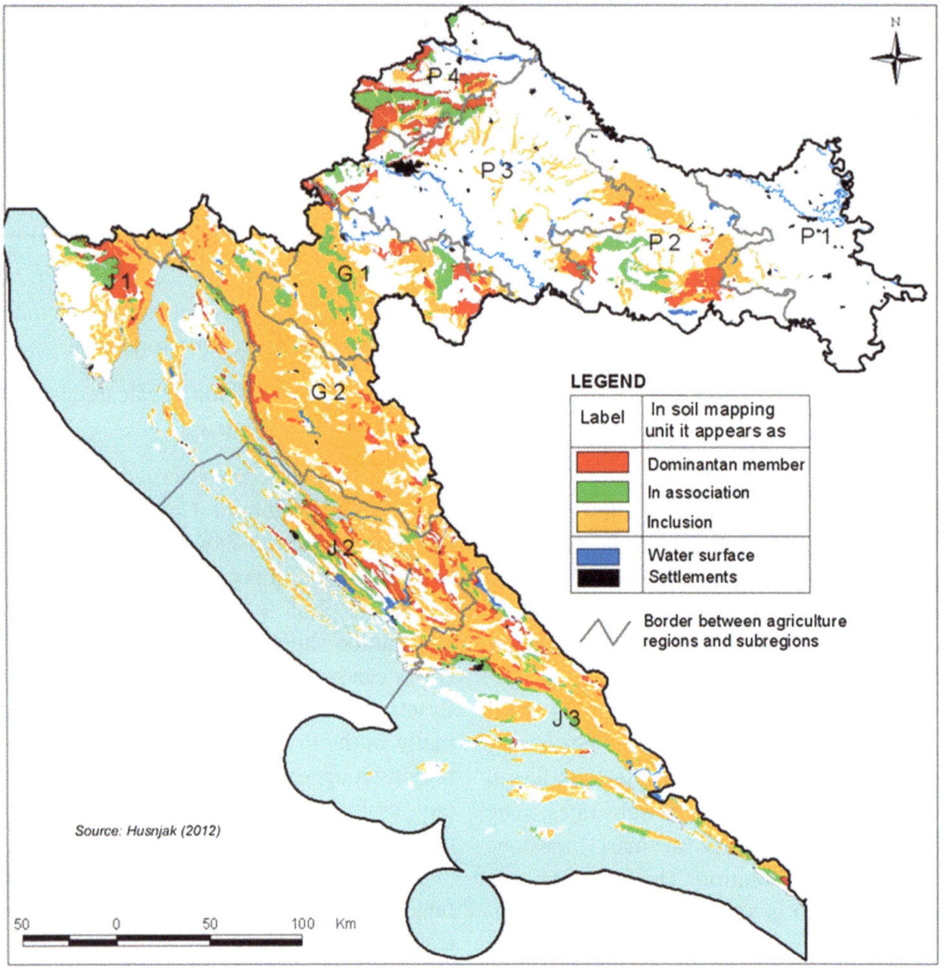

Fig. 6.9 Distribution of rendzina in the pedosphere of Croatia

Photo 6.12 Rendzina on calcareous regolith of limestone on Urljaj mountain—880 a.s.l. near Srb (*left*) and on soft, Eocene limestone in Drivenik on the edge of the Vinodol valley (*right*)

Table 6.5 Physical and chemical properties of rendzina

Soil horizon	pH (nKCl)			mg P_2O_5/100 g of soil			mg K_2O/100 g of soil			Humus (%)			Clay content (%)		
	Average	Minimum	Maximum	Average	Minimum	Maximum	Average	Minimum	Maximum	Average	Minimum	Maximum	Average	Minimum	Maximum
A	6.5	5.4	7.1	4.1	0.0	14.6	14.0	6.0	40.0	8.4	2.3	18.9	22.3	0.0	40.8
AC	6.9	5.2	7.6	2.0	0.7	4.8	8.0	1.6	13.4	2.8	0.7	5.7	23.4	0.0	49.0

Source Bašić et al. (2001)

Photo 6.13 Rendzina after erosion of soft limestone—Benkovac in the Zadar hinterland, and flysch of central Istria (Cerovje) are very fertile soils, particularly suited for growing grapes

Guidelines for management. Rendzina is a dry and warm habitat for natural xerothermic vegetation. Depending on climatic conditions and method of irrigation, their use is justified. This requires soil conservation—protection from erosion. Depending on the parent substrate, intensive fertilization is required with mineral fertilizers, especially nitrogen and phosphorus. The amount of humus should be maintained at a level of about 5 %. For this purpose, if necessary, use organic fertilizers, or practice the growing of clover–grass mixtures and green manuring. Deepening the plow layer at the expense of the transitional AC horizon or substratum on the surface can be risky for cultures sensitive to an excess of active lime such as grape, peach, orange, lemon, etc. In addition, because of the inactivation of plant-available iron in these soils iron deficiency is common—this is called ferro-chlorosis,

6.1.2.4 Chernozem

There is a local saying relating to chernozem:

> Better Christmas without meat than without snow!
> On Chernozem a button would germinate and grow.

The name of the soil type. The name is derived from the Russian language where it means "the black soil." A similar name is used in other world languages, because of the black color characteristic of this soil. Although the area of chernozem is not large it has great importance for our country, because it is one of our most valuable agricultural soils, with high and stable yields of all arable crops. Chernozem is also one of our best studied soils. All the properties of chernozem are well known, because of this soil type was investigated from all pedologists, but the most detail is studied in monograph of Škorić et al. (1977).

Genesis and sequence of horizons. Chernozem is a soil type of semiarid steppe areas with a typical molic A_{mo} horizon, deeper than 40 cm, and transitional AC horizon (25–30 cm depth) formed over loose, typically calcareous loess. The soil is calcareous from the surface, and a frequent diagnostic criterion is a krotovina or AC horizon with mould-like pseudo-spawn generated by $CaCO_3$ leaching out of the surface A_{mo} horizon and re-crystallizing on the surface of soil aggregates. The genesis of chernozem and its place in an evolutionary series on loess was first studied by Vasilj Vasiljevič Dokučajev (1846–1903), the founder of modern soil science—pedology. The genesis of chernozem is related to the specific conditions of formation: loess as the parent rock, continental climate, and steppe grass vegetation. As described in the previous chapter, the loess is a loose, calcareous substratum of loamy texture, and favorable air–water conditions; it is permeable to water. In addition, it naturally contains considerable amounts of nutrients, which are collected via biological accumulation

Photo 6.14 Typical appearance of the profile of chernozem—krotovina (*left*) and pseudo-spawn (*right*)—Laić near Dalj city in P-1 subregion (photo: Komesarović)

in the humus-accumulative A_{mo} horizon. It contains the clay mineral illite with lesser chlorite and montmorillonite. The basic processes of formation of chernozem are accumulation of mature, high-quality molic humus rich in plant nutrients—biogenic elements, creating a stable crumb structure. The age of chernozem is estimated to be between 8,000 and 10,000 years, about two or three millennia older than the Vučedol civilization. Development began in a dry and warm climate, and went through the evolutionary stage of regosol on loess. Today's climate, compared with the climate at the time of its origin, is much less favorable for recent genesis and maintenance of chernozem. The increasing amount of rainfall reduced the mean annual temperature and the typical steppe vegetation due to climate change. The chernozem zone in Croatia is transitional, semiarid to semihumid, on the boundary between a typical steppe zone and zone of forest-steppe. Aridity of the climate is much less than in the typical recent chernozem zone in Russia. Such climatic conditions are the cause of weak chemical weathering and leaching of the substrate and the appearance of special steppe vegetation dominated by grasses. They create a huge overhead mass, and a very dense root system permeates the surface layer of the soil. Paralysis of vegetation at the end of the vegetation period each year accumulates humus of very high quality, saturated with bases, and rich in all the biogenic elements. The biological activity of chernozem is high, but in the process of mineralization there are two interruptions, the first due to drought in summer, and the second in winter due to low temperatures. Therefore, the balance in the transformation of organic substances is a positive one on the side of accumulation of humus. In chernozem formation, soil fauna plays an important role—earthworms and steppe rodents (ground squirrel, hamster). Earthworm excrement and krotovina are normally present in the soil profile (Photo 6.14).

Table 6.6 Classification of chernozem

Name of taxon	
Variety	Form (based on depth to loess)
Typical (calcareous)	Shallow 40–60 cm
Leached (non-calcareous)	Medium deep 60–80 cm
Cambic (initial $(B)_o$ horizon)	Deep > 80 cm

Chernozem is a soil of lowland areas of Croatia, on loess plateaus (90–140 m a.s.l.). As eolian formations, plains on loess plateau are not ideal plains, as they show variable *mezzo* elevations and *meso* depressions. Depressions may be affected by high groundwater, flushing, and increased accumulation of material eroded from the tops of dunes. The chernozem is strongly influenced and affected by human destruction of natural vegetation and cultivation, causing a significant adverse change in chernozem: decline of humus content and nutrients, compaction, structure deterioration, tillage erosion, and general decrease of soil fertility, which is naturally very high.

The WRB name. Chernozem

Position in an evolutionary series. Chernozem is a member of an evolutionary series which starts with regosol on loess and contains the following soil units:

Chernozem (typical-calcareous → leached → cambic) → **Eutric cambisol on loess → luvisol**...

The taxonomy of chernozem on the lower level is presented in Table 6.6.

Calcareous chernozem is a typical chernozem, the most famous, most investigated and best known soil and the most fertile soil in the world.

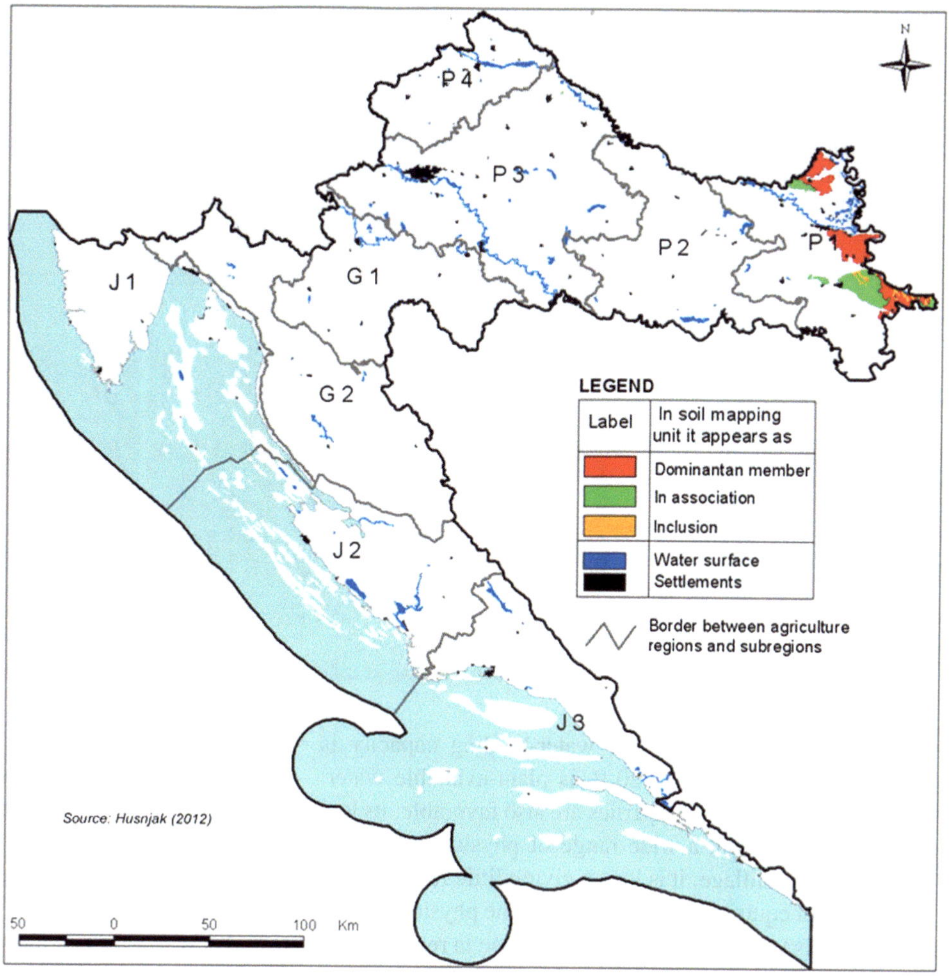

Fig. 6.10 Distribution of chernozem in the pedosphere of Croatia

Leached chernozem is soil in which leached $CaCO_3$ from the A_{mo} horizon is washed out and accumulates in a transitional AC horizon in the form of pseudo-spawn. It occurs in mild depressions, with increased lateral wetting as the first stage of degradation of chernozem in west Syrmium (eastern Slavonian P-1).

Cambic chernozem is the next stage of degradation—the evolution of chernozem soil into a mature stage. The morphology is typical: the A_{mo} horizon is shallower, becomes dark gray because of the mineralization of humus, and $CaCO_3$ is washed out to the substrate, where it forms "loess dolls." The initial $(B)_o$ horizon is thin but visible, as a sign of the increased chemical weathering process—leaching, decomposition of primary minerals and synthesis of secondary minerals—clay formation—initial argilogenesis.

The evolution of chernozem takes place in the direction of forming cambisol eutric under the influence of increased rainfall and altered forest vegetation.

Distribution. Croatia has 51,808 ha of chernozem, or 0.93 % of the land area. It occurs in the eastern Pannonian subregion P-1, in west Syrmium and Baranja, then on a narrow strip along the Danube, around Erdut, further around Vukovar (Bobota, Vera, and Klis), Tovarnik, Šarengrad, and Ilok (Bogunović et al 1984; Šalinović et al 1999; Bogunović and Husnjak 1999). It is usually a marginal, peripheral zone in which the zone of cambic soils begins (Fig. 6.10).

It regularly occurs in association with the "younger" members of the series of soils on loess, primarily with regosol, generated by soil tillage and the tillage-erosion of tilled soil on the wavy meso-relief of loess plateaus. It occurs as inclusions with cambisol eutric and luvisol on loess in the western part of the loess plateaus between Vukovar and Mohovo Opatovac and Lovas and Tovarnik (Photo 6.15).

Physical and chemical properties. These are very favorable for plant growing. The physical properties are shown in Table 6.7.

Chernozem is a soil of loamy texture, with clay content of 20–30 %, which increases under the influence of cambic processes. In the clay fraction, clay minerals of the illite group dominate. The structure is favorable—crumbly, which is optimal for a favorable air–water relationship. Total porosity is about 45 %, the relationship between micro- and

Photo 6.15 *Left* The typical chernozem: a very deep molic horizon followed by an intermediate A/C, which diffuses into the substrate—loess (photo: Husnjak). *Right* On the hill is regosol as the consequence of tillage erosion

macropores is optimal, at 2:1, water-holding capacity is 32–34 vol% of which over 20 % is plant-available water. Texture and all physical properties are also favorable: its low density and adhesion, a wide range of physical properties favorable to soil tillage; it is loose, giving little resistance to penetration of equipment for soil tillage. The physiologically active layer is very deep; there is no increase in resistance to root penetration to deeper horizons. Chemical properties are also favorable (Table 6.8).

The chemistry is practically neutral to slightly alkaline. It has a medium nutrient supply, humus content is quite low with a tendency to decrease further because intensive tillage increases mineralization of humus as a typical aerobic microbiological process. The humus is of high quality, saturated with bases, with a low, practically optimal C:N ratio (about 10). The soil is calcareous throughout the whole depth profile. The CEC is 30–35 meq/100 g soil, which is significantly lower than in the Russian chernozem, due to the lower content of humus and the domination of illitic clay minerals. Chernozem shows very high biological activity, and is rich in microorganisms, especially bacteria, including symbiotic and nonsymbiotic nitrogen-fixing bacteria. Chernozem contains numerous microfauna, especially earthworms.

Land use. Chernozem is used as arable land for intensive cultivation of field crops. It is a naturally fertile soil due to the optimal physical, chemical, and biological properties described. It provides the widest range of crops, and yields are high, although in dry years without irrigation systems they can fall significantly below the usual high average. The most famous and oldest farm—the over 310-year-old Belje farm—was extremely successful and profitable thanks to chernozem, as a soil type on which it is necessary "not to forget to sow crops" (Photo 6.16).

Guidelines for management. The local saying printed below the title of this section illustrates the main limiting factor of farming on these soils—it is water-deficient. It means that if Christmas is without meat but with snow, the next harvest will be rich. With a changeable and unpredictable climate, irrigation is obligatory to achieve high and stable yields (Bašić and Bašić 2007). Also, the only limit to higher yields is water deficiency. Irrigation has to be accompanied by change—intensification of arable farming activities.

Chernozem is a soil type with relatively high overall nutrient content, and thus in this respect has a high potential fertility. However, effective fertility is lower; nitrogen is bound in the humus and only slowly released by mineralization. Nitrogen nutrition from these reserves meets production requirements only a few years after starting cultivation, and to maintain the strict requirement of high and stable yields on chernozem it is necessary to practice fertilization with nitrogen fertilizers. Total content of phosphorus is also respectable, but a significant part is chemically fixed in tertiary calcium phosphate, or in organic substances that are released slowly and in amounts that are insufficient for the high yields expected on chernozem. All arable crops show a very positive response to phosphorus fertilization. Chernozem has a naturally high content of potassium, but also a great possibility of it being fixed in the interlamellar space of illite clay minerals. However, potassium fertilization in relation to N and P on chernozem shows a much smaller effect.

Table 6.7 Physical properties of chernozem

Soil horizon	Clay particles (%)			Total porosity (vol.%)			Water-holding capacity (%)			Air cap. (%)			Plant-available water (%)		
	Average	Minimum	Maximum	Average	Minimum	Maximum	Average	Minimum	Maximum	Average	Minimum	Maximum	Average	Minimum	Maximum
A_p	24.7	20.6	29.8	45.6	45.3	45.9	33.0	32.2	33.9	12.5	11.4	13.7	20.4	19.7	21.2
AC	23.5	14.6	26.9	48.2	47.9	48.5	34.9	32.6	37.2	13.3	10.7	15.9	25.5	22.0	29.1
C	21.1	18.4	26.7	49.1	47.6	50.6	35.0	32.7	37.4	14.0	10.2	17.9	–	–	–

Source Škorić et al. (1977)

Table 6.8 Chemical properties of chernozem

Soil horizon	pH in H_2O			P_2O_5 mg/100 g of soil			K_2O mg/100 g of soil			Humus (%)			$CaCO_3$ (%)		
	Average	Minimum	Maximum	Average	Minimum	Maximum	Average	Minimum	Maximum	Average	Minimum	Maximum	Average	Minimum	Maximum
A_p	7.5	7.1	8.2	9.7	1.3	21.8	16.2	10.6	28.0	2.0	1.5	2.4	22.8	0.4	79.7
AC	7.9	7.2	8.4	2.1	1.2	6.1	7.8	4.4	15.0	1.3	1.0	1.8	41.6	8.2	59.3
C	8.0	7.5	8.4	–	–	–	–	–	–	–	–	–	46.5	6.3	84.1

Source Škorić et al. (1977)

Photo 6.16 View towards "the largest renewable treasure," as part of the "green Pannonian sea"—the "endless" arable farm "Belje" managed on chernozem and humogley. It has for over three centuries achieved the highest arable crop yields, but chernozem is "thirsty"! The next step in the never-ending story of progress on "Belje" is irrigation of every square meter of the soil! Chernozem is a grateful soil just for irrigation!

One of the important measures for maintaining the high natural fertility of chernozems is the choice of proper crop rotation and the sowing of clover–grass mixtures in the rotation. With chernozem having the highest natural fertility in the world (especially in the U.S.) for some crops, soil tillage is gradually being abandoned and the so-called minimum or no-tillage approach is being introduced, with water conservation measures through a natural mulch (straw and cornstalks to cover the soil surface). Under the conditions present in Croatia, processing experiments showed full justification for the use of such processes (Butorac 1993; Jug et al. 2010; Birkas et al. 2008). Even though new machinery would need to be purchased, the future definitely lies in this direction.

6.1.2.5 Vertisol

The name of the soil type. The name is derived from the Latin word *vertere*—rotate, due to the process of "self-rotation"—pedoturbation—that occurs in this soil. In Croatia we use the Serbian name Smonitza (from *smola*—resin), which is accepted in other countries in south-east Europe and was given by the famous Serbian soil scientist Stebut (1949, 1950, 1953), who first studied the genesis of vertisol in Serbia. The origin of the name illustrates the consistency of this soil. In agronomic circles vertisol is called the "minute soil," which means that it is necessary to till this soil only once—every minute is important. If it is dry it is impossible for a plow to penetrate this soil; if it is wet it is plastic and can hamper tillage operations. The population of Istria use the Italian name terra nera (black soil) for this soil.

Genesis and sequence of horizons. Vertisol is a soil type of A–AC–C soil profile horizon sequence, very similar to chernozem, but formed on substrates originally rich in clay with a high proportion of clay minerals of the montmorillonite type. There is a deep humus horizon in which, because of the low permeability of the clay, rainwater often stagnates and there are some signs of hydromorphic processes. The molic humus-accumulative horizon shows some characteristics of aquatic processes; it is rich in humus with a characteristic strong dark color. The process of pedoturbation leads to a characteristic surface relief, known as "džomba," where the soil has a rugged, bumpy microrelief.

The WRB name. Vertisol

Position in an evolutionary series. Vertisol is generated on clay sediments of Tertiary age containing $CaCO_3$. The substrate "inherits" large amounts of montmorillonitic clay which plays a significant role in the genesis and characteristics of vertisol. In this regard vertisol is an extremely lithogenous soil. It occurs on all forms of relief—on flattened lake terraces and in gently undulating positions with semiarid growing conditions, a mild climate with warm and dry periods. The alternating cycles of contraction when dry and swelling when wet causes the phenomenon called pedoturbation. How does this occur? In a dry period, drought causes evaporation of water and contraction of volume which forms cracks in the soil. When it rains the cracks fill with rainwater, causing dispersion and suspension of clay particles. The water molecules penetrate through hydration into the interlamellar space of the montmorillonite clay, causing expansion of the clay in the cracks and leading to closure of the cracks. The effect of this expansion is volume and weight increase and the soil rises creating an uneven (bumpy) surface microrelief. The expansion forces lead to the formation of "slickensides" or slip plains (shiny surfaces caused by movement of soil blocks due to expansion pressure) in the soil's surface layer. Stebut considered Smonitza relict soils, developed in the lake and postlacustral (postlake) stage (Tertiary and Post-Tertiary period). The lake stage developed lush wetland vegetation whose remains transformed and fermented under special anaerobic conditions in the presence of large amounts of clay minerals. The lakes later receded, and Smonitza became the surface under the new terrestrial conditions. The A_{mo} horizon of the soil contains some "conserved" aquatic signs of the soil's previous existence in the shallow lake. Under recent conditions vertisol has the character of an automorphic–terrestrial soil, so the evolution taking place in Smonitza is under automorphic conditions. Because of its low permeability the leaching of vertisol is a very slow process, which continues in the following evolutionary series:

Table 6.9 Taxonomy of vertisol

Name of taxon	
Variety (based on process)	Form (based on depth)
Calcareous	Shallow 30–40 cm
Non-calcareous (leached)	Medium deep 40–60 cm
Cambic	Deep > 60 cm

Vertisol (leached → cambic) → **Cambisol eutric**, on lake sediments

The evolutionary path of vertisol is also visible from its systematization at the level below subtype (Table 6.9).

Distribution. It occurs over a small area of only 5,002 ha, less than 1 % of the Croatian land area. The heavy, Tertiary (Miocene or Mio-Pliocene) vertic clay soils on the border with Bosnia (at Velika Kladusa) are similar to vertisol but without a humus-rich A horizon. We found these clay soils near Perjasica and Tounj and Vojnić in Kordun (Fig. 6.11).

During our own research we found larger areas of vertisol, terra nera in the area of Buje, Brtonigla in western Istria (Bašić and Adam 1979), where these soils are locally distributed on soft marl limestone of Miocene age. The remarkable feature of these soils is their black color, which is why the population call this soil terra nera. It is used for growing grapevines. As terra nera is a fresher, cooler soil, with high water-holding capacity, all sorts of grape growing on this soil provide fresher wines, with a higher acid content. The taste of Terran, a famous Istrian red wine, has an unforgettable extra quality, more acid than on surrounding soils on flysch and much more than on terra rossa.

Physical and chemical properties. Vertisol is a heavy, clayey soil often with over 40 % of clay, mainly (90 %) of montmorillonite type. It has adverse physical properties: a high water-holding capacity of which only a small part is plant-available, low water permeability, high plasticity, adhesion, high resistance to all tillage operations, and a short period of physical maturity for tillage (therefore known as the "minute" soil). In the dry state, it is hard, creating cracks and fractures. In this situation, upon the first fall of rain it is permeable to water, leading to a decrease in air content in the soil. Vertisol has a neutral to slightly alkaline (calcareous) or slightly acid (cambic) chemistry. Humus content is 3–5 %, and the total amount, due to the depth of the A_{mo} horizon, is very large. Humus is associated with montmorillonite in a very stable organo-mineral complex. It has a high CEC and a high saturation of colloidal complexes by bases. In terms of biogenic elements it shows a lack of available phosphorus (Photo 6.17).

Land use. Vertisol is a highly valuable agricultural soil, with high potential fertility, despite the characteristics described above. This soil is mostly used as arable land,

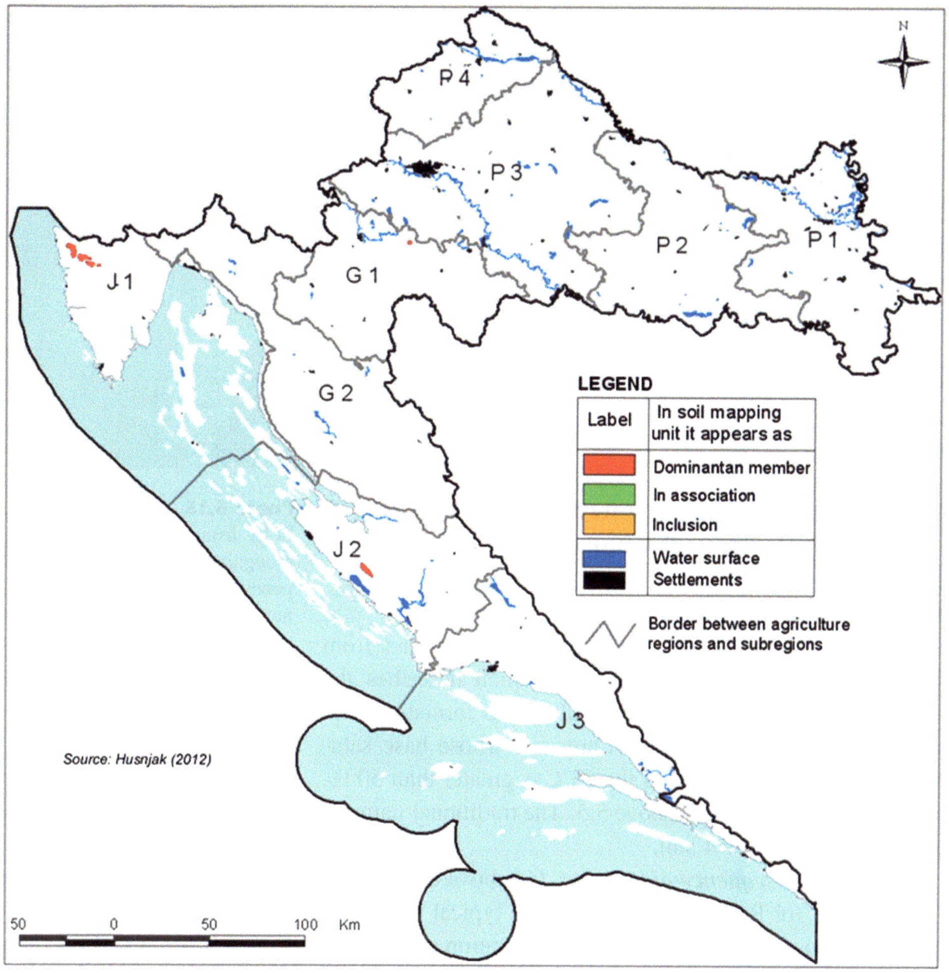

Fig. 6.11 Distribution of vertisol in the pedosphere of Croatia

Photo 6.17 In dry condition vertisol forms deep, vertical cracks in the soil and on the surface

vineyards, and orchards. It provides high and special quality to wine. In Istria it is very favorable for the durum type of wheat. If it is to be used for vegetable growing, this soil needs irrigation and deep winter plowing to expose it to frost for a favorable soil structure.

Guidelines for management. Vertisol's use as arable land is very important, and deep autumn plowing and exposure to winter frosts is recommended as an essential measure, which provides a stable soil structure. Vertisol requires irrigation to provide secure and high yields of arable and vegetable crops in particular. In addition, if intensive production is practiced this soil needs fertilization, with an accent on phosphorus.

Photo 6.18 Indeed, if we were to find this paleosol (Šarengrad near Ilok) from the last Wurm on the surface of a soil it would be recognized as a $(B)_o$ horizon of recent soil genesis. If recent soil genesis is resulting in the same profile as what was "conserved" by a new glacial period, is our interglacial also coming to an end? Our generation may view this paleosol as simply a "memory," but that does not mean that nothing new will occur on the Earth! There is no place for anthropocentrism!

6.1.3 Class III: Cambic Soils A–$(B)_o$–C, or A–$(B)_o$–R Soil Horizons Sequence

6.1.3.1 Cambisol Eutric

The name of the soil type. The name cambisol comes from the Italian word *cambiare*—change, which describes the character of the processes that result in the formation of a $(B)_o$ horizon, and *eutric* means any soil whose base saturation (with basic cations) of the CEC is greater than 50 %, and whose pH in water is above 5.5. The traditional name of this soil type is brown soil.

Genesis and sequence of horizons. Cambisol eutric has a A_{mo}–$(B)_o$–C (or R) profile sequence. The typical horizon is a cambic $(B)_o$ in which the degree of saturation of the CEC is greater than 50 %, and the chemistry in water is over pH 5.5. This soil type is formed on calcareous or basic saturated parent substrates like loess, other sediments rich in bases, neutral and basic eruptive or metamorphic rocks. Rainfall in the zone of cambisol eutric ranges from 600 to 650 mm, i.e., slightly more than in the chernozem zone. The mean annual temperature ranges from 10 to 12 °C. It is generated at altitudes up to ca. 200 m, on flat and slightly undulating relief forms.

The process of formation is complex and lengthy; according to our calculations, on loess of west Syrmium it takes around 7,000–8,000 years. If so, taking into account the morphology of paleosol on loess, which is very similar to the $(B)_o$ horizon of cambisol eutric, we are very close to the conclusion that a new (or the next) glacial period is not very far away! This means that although we are continuously discussing (global?) warming, (global) glaciations are *ante portas*! (Photo 6.18)

The cambic process includes several basic processes of pedogenesis. The first is leaching of $CaCO_3$ and decarbonatization of the upper part of the profile, caused by increased rainfall compared with that typical for chernozem. Arriving at a lower soil pH—low acidification, the mild and wet winter and spring enable enhanced mineralization of humus, and its content therefore gradually declines. Regular tillage and aeration from the time of the Vučedol civilization until today also support the process of mineralization and decreasing humus content. The roots of forest vegetation normally excrete CO_2, which, after acidification and leaching of carbonates, supports the processes of chemical decomposition of primary minerals and the formation of secondary minerals during the process of argilogenesis. This process causes slight textural differentiation of profiles. The content of clay in the $(B)_o$ horizon is slightly higher than the content in the A_{mo} horizon. The relationship between SiO_2:R_2O_3 is constant throughout the depth of the profile, which signifies the absence of sesquioxide migration by leaching. The parts of the iron oxides that are not synthesized to secondary clay minerals in the $(B)_o$ horizon give it its brownish color.

The WRB name. Eutric cambisol

Position in an evolutionary series. Eutric cambisol on loess is formed as a zonal soil from chernozem cambic, which means that the zone of this soil naturally continues on from the zone of chernozem. At the beginning of argilogenesis the initial cambic $(B)_o$ horizon is thin and appears as a variety of chernozem cambic on loess but on other parent substrates as rendzina cambic in the form of narrow strips of brown color. Continuing to evolve, this initial horizon gradually expands toward the surface and the deeper layers,

Table 6.10 Taxonomy of cambisol eutric

Name of taxon	
Variety (based on process of genesis)	Form (based on texture)
For all subtypes • Typical • Luvic • Pseudogleyic	For all varieties • Sandy • Loamy • Clay (vertic)

When its depth exceeds the depth of the A_{mo} horizon, the soil passes from the class of humus-accumulative A–C to the class of cambic A_{mo}–$(B)_o$–C soils. The evolution of eutric cambisol to the next evolutionary stage begins with an initial eluvial–illuvial clay migration process, followed by an initial textural differentiation and formation of E and B_t horizons.

Outside of the loess zone it appears as a member of an evolutionary series of soils on substrates rich in basic cations. The relief and natural vegetation influence the direction of this soil type formation—in humid areas correcting the condition to xerothermic but in the arid to more fresh, wetter conditions (Table 6.10).

Distribution. The distribution of cambisol eutric is quite high, occupying 172,495 ha or 3.1 % of the Croatian land area. This soil is found as a zonal type and as the dominant soil type in cartographic units on the slopes of the Fruska gora hills in Ilok, on the Vukovar loess plateau, around Vukovar and Vinkovci. It usually occurs in a soil association with regosol, chernozem, and luvisol on loess. In addition to the P-1 subregion, it occurs in P-2 on leached loess, and P-3 and P-4 on base-rich marl substrates, including multilayer fluvial deposits of the Sava and Drava rivers (Photo 6.19, Fig. 6.12).

Physical and chemical properties. These depend on the character of the parent substrate C or R type and its chemical properties, which influence the types of clay mineral and duration of argilogenesis. Some authors discuss the influence of vegetation, especially that of forests, the rhizosphere of which causes cambic processes by respiration from the roots, emission of CO_2, and slight acidification (Bogunović et al. 1984) (Table 6.11).

As can be seen, the clay content in $(B)_o$ is increased. Under natural vegetation (forest) this soil has a stable, crumbly structure and favorable air–water relationship, but on cultivated land it is somewhat worse. However, eutric cambisol is a well drained and warm soil. The content of humus, its quality, and thickness of the humus-accumulative A_{mo} horizon depend on other bio-climatic conditions and form of use. In the fields, the average content of humus is 2–3 %, in the forest up to 7 %. However, in forest the A_{mo} horizon is much shallower. The chemistry is slightly acid to almost neutral, somewhat lower in the $(B)_o$ horizon. It has a high CEC, slightly higher in the $(B)_o$, horizon and the degree of saturation by bases is higher than 50 %: usually above 80 %. In the cation exchange (CE) colloidal complex Ca^{++} and Mg^{++} ions dominate.

Land use. The natural vegetation in this zone is transitional forest-steppe; most has been cleared and cultivated but rests of forest are aesthetically and economic valuable oak (*Quercus rubra*) of the forest association *Carpino*

Photo 6.19 *Left* Typical profile of eutric cambisol on loess (photo: Husnjak). *Right* The plowed surface of eutric cambisol on the typical eolian meso-relief of Vukovar loess plateau. We see "dunes" with a $(B)_o$ horizon of brownish color at the surface, in association with regosol (*yellow color*) on calcareous loess. At the bottom of the eolian depressions there is intensive leaching and an initial eluvial–illuvial migration process

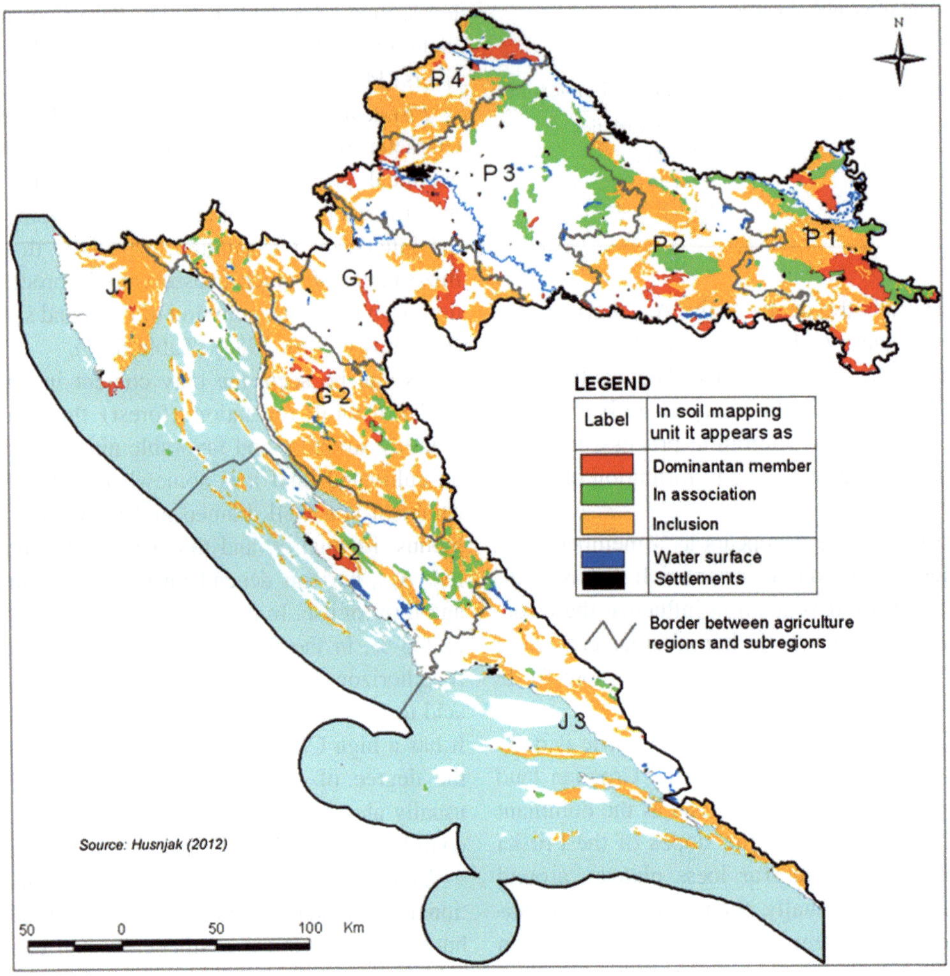

Fig. 6.12 Distribution of cambisol eutric in the pedosphere of Croatia

Table 6.11 Texture and some chemical properties of eutric cambisol on loess

Soil horizon	pH in H_2O			P_2O_5 mg/100 g of soil			K_2O mg/100 g of soil			Humus (%)			Clay content (%)		
	Average	Minimum	Maximum	Average	Minimum	Maximum	Average	Minimum	Maximum	Average	Minimum	Maximum	Average	Minimum	Maximum
A_p	6.6	5.3	7.5	6.5	0.4	24.2	18.6	2.8	34.0	2.3	1.0	7.3	26.4	18.7	40.2
$(B)_o$	6.6	5.1	7.7	4.4	0.3	9.1	8.7	2.8	15.8	1.1	0.6	4.7	30.6	20.1	50.8
C	7.8	6.8	8.4	–	–	–	–	–	–	–	–	–	27.4	18.2	59.8

Source Škorić et al. (1977)

betuli—Quercetum roboris, and there is also lime-tree (*Tillia grandifolia*), ash (*Fraxinus excelsior*), and acacia (*R. pseudoaccacia*).

All properties of eutric cambisol on loess are very favorable for agricultural and forest production. In fertility it is comparable with chernozem, and because of the greater rainfall it gives quite high and stable yields of agricultural crops. The choice of crops on this soil is wide—it is suitable for arable, vegetable crops, cultivation of tobacco, fruit and vineyards. The most famous vineyards of Ilok and Baranja are only on eutric cambisol and regosol on loess (Photo 6.20).

Our research shows that, should the paleosol crop out on the surface and mix with regosol or loess, it influences the quality of the grape and wine. Galović et al. (2008) suggest that the sequence of loess deposition and the succession of the individual layers represent remarkable "archival material" and provide evidence for past environmental and climate change in the Middle and Late Pleistocene. They found at least six layers of paleosols (meaning six interglacial periods!) within the deep loess deposits. It is our opinion that paleosol, when and if it is found in the rhizospheric zone or on the surface, gives a special, unique

Photo 6.20 *Left* The oldest and most famous vineyards of the farm "Belje"—Agrokor wine, the pride of Belje, in Baranja hill on eutric cambisol (in the foreground) in association with regosol on loess (light zones in inter-row space). *Right* The same association of soils under grapes grown in Ilok, from which arises the "Prince of Principovac"—Traminer

quality to the sites on which vines in Ilok and Baranja are grown. This issue should be explored by studying the number of paleosols and the depth at which they occur, the penetration capabilities of prospective grapevine roots to these layers, and their benefits for growth and root development. In some cases this is undoubtedly on the soil surface.

Guidelines for management. Properly chosen and timely executed, tillage and other plant-growing activities on eutric cambisol on loess are the first precondition for exploitation of the high potential of these soils. This soil is very erodible if exposed to heavy torrential summer rains. This means that the first measure must be protection against erosion on sloping terrain. On fields with a steep slope it is recommended that the rows and/or grass cover of inter-row spaces in the plantation are oriented with the contours. On arable land, high and stable yields require fertilization of N, P, and K fertilizers. Particularly favorable are the effects of nitrogen fertilization. Because there is no risk of fixation or leaching, fertilization can be performed using high, so-called amelioration doses; recommended is humification, deep autumn tillage for row crops such as sugar beet, and proper crop rotation including clover with grass mixtures. The presence of this soil type ensures successful farming.

Eutric cambisol on other parent material is generally speaking a very good soil, but this depends on other conditions, like perhaps a high groundwater level, or limits to root penetration if the parent rock is of R type. Because forests on this soil are reliquiae reliquiarum—"relicts of relicts," they create extremely beautiful landscapes, full of big and small game, and mushrooms, but they are permanently imperiled by different space consumers and radical changes in water regime, and therefore it is necessary to protect these natural beauties.

6.1.3.2 Cambisol Dystric

The name of the soil type. The term *dystric* refers to any degree of saturation of the soil CEC of less than 50 % and pH values in water less than 5.5. In older literature this is described as *brown acidic soil.*

Genesis and sequence of horizons. The genesis of this soil is highly correlated with the properties of the parent substrate, which is poor in bases, and therefore acidic, silicate rocks—igneous, metamorphic and sedimentary rarely, compact or loose (granite, gneiss, phyllites, shales, sandstones, and cherts). Because of the influence of the parent rock on genesis and all properties of the soil, these soils are typically lithogenic soils. The dominant process of genesis is chemical destruction—decomposition of primary minerals and argilogenesis. The lack of bases leads to the lack of larger amounts of clay minerals. Unfavorable climatic conditions and high rainfall leads to accumulation of humus (because of low mineralization), leaching, and acidification. The soil profile has the sequence of soil horizons: A_{oh}–(or A_{um})–(B)–C (or R), with low saturation of the CEC (lower than 50 %) and a soil chemistry in water below pH 5.5.

Table 6.12 Taxonomy of dystric cambisol

Name of taxon	
Variety (based on process)	Form (based on soil depth)
Typical—no signs except for the typical cambic process	Shallow < 40 cm
Humus rich—accumulation of humus on the surface	Moderately deep 40–80 cm
Luvic—initial E horizon of luvisol	
Pseudogleyic—initial signs of pseudogleyic processes	Deep > 80 cm
Podzolic—initial E spodic horizon	

The WRB name. Cambisol dystric

Position in an evolutionary series. The evolution of this soil type flows in two directions. Parent substrates originally richer in terms of base content provide a little more clay, and the evolution of the soil exceeds eluvial–illuvial migration of clay minerals. But on porous, permeable substrates very acid poor in terms of bases, this soil is subject to podzolization processes, i.e., evolution in the direction of formation of podzol. The first stage of this process is cambisol dystric, podzolic, followed by podzols by the complex of processes described. It is also visible in the taxonomy at the level of variety and form (Table 6.12).

Distribution. Dystric cambisol occupies 316,184 ha or 5.7 % of the Croatian land surface (Fig. 6.13).

The dominant unit is on acid substrates of eruptive and metamorphic origin in the Middle Slavonian hills in P-2, Moslavacka hill and part of Papuk in P-3, Medvednica, Macelj, and Ivančica in P-4, and Petrova and Zrinska hill, part of Velika and Mala Kapela and Gorski Kotar in the G-1 and G-2 subregions. The association occurs in the same areas, and it is a widespread soil as inclusions (Photo 6.21).

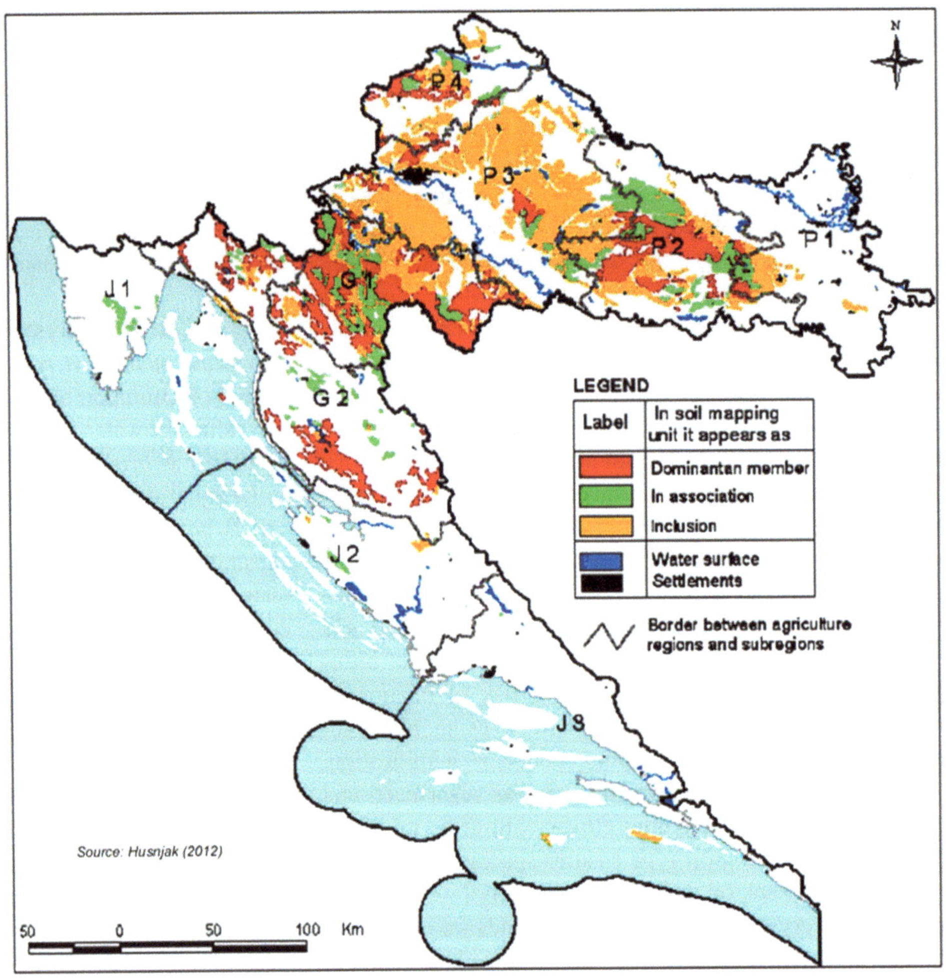

Fig. 6.13 Distribution of cambisol dystric in the pedosphere of Croatia

Photo 6.21 Dystric cambisol on sandstones—Dedin near Zalesina (*left*) in Gorski kotar—typical habitat of acidophil Blue Berry (*Vaccinum myrtillus*), G-2 subregion (*right*)

6.1.4 Class IV: Polygenetic Cambisols of Limestone—Dolomite Karst of A–$(B)_{no}$–R Sequence of Soil Horizons

There are several reasons for identifying a special class of polygenetic soils on karst.

In the first place there is the fact that in the WRB soil taxonomy there is no logical "place" for these soil types. The genesis of soil types of this class occurs over a very long time period in which the bio-climatic conditions changed. The main, specialized, and very slow process of genesis is dissolution of limestone and/or dolomites and accumulation of clay minerals as an insoluble residue. This process is very different from argilogenesis in the cambic process of all other cambic soils, defined as the decomposition of primary and synthesis of secondary (clay) minerals. Soils of this class are far older than cambic soils, and the colors of the genetic horizons are very different.

The second reason is the alleged difference and overlap between the two types of soil in this class, kalkocambisol and Rhodic soil (terra rossa). Many of our researchers, from geologists Kramberger and Tućan, through Gračanin (1951) and Škorić et al. (1987), including personal surveys (Bašić and Adam 1979) and the recent publications of Durn (2003) with results obtained using the most modern laboratory equipment and methods of analysis, elucidated the genesis of red soil as a process of dissolution of limestone and accumulation of insoluble mineral residues which give the red color. It was red in color at the time of the formation of the limestone. Today these soils are also being generated, but under recent conditions of the Mediterranean climate they retain their original characteristics. Yet under different conditions in continental Croatia the soil is transformed by cambic processes evolving in the direction of eluvial–illuvial migration of clay minerals. Keeping in mind that these soils have to find their own logical position in every global soil taxonomy, we want to provide our own contribution to this question.

There is still another, no less important reason: these soils are of great prevalence, high fertility and economic importance, especially in the Adriatic region including the islands as well as in the whole Mediterran.

6.1.4.1 Kalkocambisol

The name of the soil type. The name originates from the term used in the WRB classification: the German word *Kalk*—lime, which suggests a soil formed on limestone and dolomites, and *cambisol* (*cambio*—change). In the past it was described under the name brown soil on limestone, also in German literature as terra fusca. The soil has a characteristic color which can be yellowish to reddish and has a very stable, markedly sharp-edged polyhedral structure. The soil is formed by dissolving limestone and/or dolomite and by accumulation of insoluble residue on stable relief positions protected from water and wind erosion.

Genesis and sequence of horizons. The specifics of pedogenesis on limestone have been described and elucidated by numerous researchers. We have experience of this soil from numerous sites in the Mountain and Adriatic regions, including the islands. Fundamentally, there is a process of dissolution of limestone occurs and the accumulation of an insoluble residue that forms the mineral component of the soil, or the $(B)_{no}$ horizon. It is formed on flatter (less steep) forms of relief under the natural vegetation of a degraded karst forest, maquis, garrigue and rocky pastures. The sequence of horizons in the profile is different and depends on the nature of the substrate: A_{mo}–$(B)_{no}$–R or C/R. Of course, if anthropogenized there is also a p subhorizon or P horizon.

Table 6.13 Taxonomy of kalkocambisol

Name of taxon	
Variety (based on depth of soil (cm))	Form (based on soil texture)
<35 cm (shallow)	Loamy
35–70 cm (moderately deep)	Clayey–loamy
>70 cm (deep)	Clayey

Anthropogenic influence is long-term and very radical, first consisting of removing stones, then building dry walls for protection from eolian and water erosion.

The WRB name. No adequate name!

Position in an evolutionary series. Kalkocambisol is a member of the evolutionary series of soils described on limestone and dolomites. It arises from kalkomelanosol cambic and evolves as follows:

KALKOMELANOSOL cambic (increased mineralization of humus and the accumulation of insoluble residue) → **KALKOCAMBISOL, typical** (deep mineral component – insoluble residue) → **luvic** (decreased humus content, differentiation of clay content in profile, E horizon formation) → **LUVISOL**

The division into subtypes is done on the basis of the dominant processes with *typical* if there are no signs of other processes and *luvic* if there is a clearly visible eluvial E horizon. Division at the level lower than subtype is shown in Table 6.13.

In nature the shallow varieties can be found on positions exposed to erosion in which water and wind translocate soil material, and the deep variety we find on topographic positions that allow sedimentation and accumulation of eroded material at the foot of slopes or in depressions such as karst holes and fields. These varieties are generally skeletal, but there are also exceptions.

Distribution. In the Croatian pedosphere, kalkokambisol covers an area of 474,959 ha or 8.53 % of the territory. Occurring on limestone and dolomites of the Alpine and Dinaric mountain system, kalkocambisol is the most widespread soil type of the Croatian karst. As the prevalent soil unit or in association with other soils of the evolutionary series of soils on limestone (lithosol, kalkomelanosol, rendzina, and luvisol) it occurs in the entire Adriatic agricultural region, including the islands and the G-2 subregion, while in the G-1 subregion and all subregions of the Pannonian region it occurs only as inclusions, except on the tops and crests of the Middle Slavonian hills, and on the crests of Ivančica, Macelj, kalnik, where it occurs in small areas as the dominant unit (Photo 6.22, Fig. 6.14).

Physical and chemical properties. Kalkocambisol is usually shallow soil, with a loamy or heavier texture and a high content of kaolin clay, which can reach up to 80 %. However, its very distinct, stable, and polyhedral structure enables very favorable air–water relationships, infiltration of rainfall water, and good permeability for water.

Although formed in limestone, kalkocambisol is a non-calcareous soil. Its chemistry is slightly acidic but can range from strongly to weakly acidic, with a high degree of base saturation (80–90 %), which is lower in the A horizon (Table 6.14).

All soil types generated from limestone kalkocambisol are poorly supplied with plant-available phosphorus, but the potassium supply is much better. In cultivated soils of this type humus content ranges from 2 to 4 %, but in forest this

Photo 6.22 Typical shallow kalkocambisol on the plain—Pernat village on the island of Cres (*left*). A stump of Quercus pubescens has remained "embedded" in the dry wall as a mute witness to the diligence of the Istrian people—Antončići at Žminj—Central Istria (A-1)

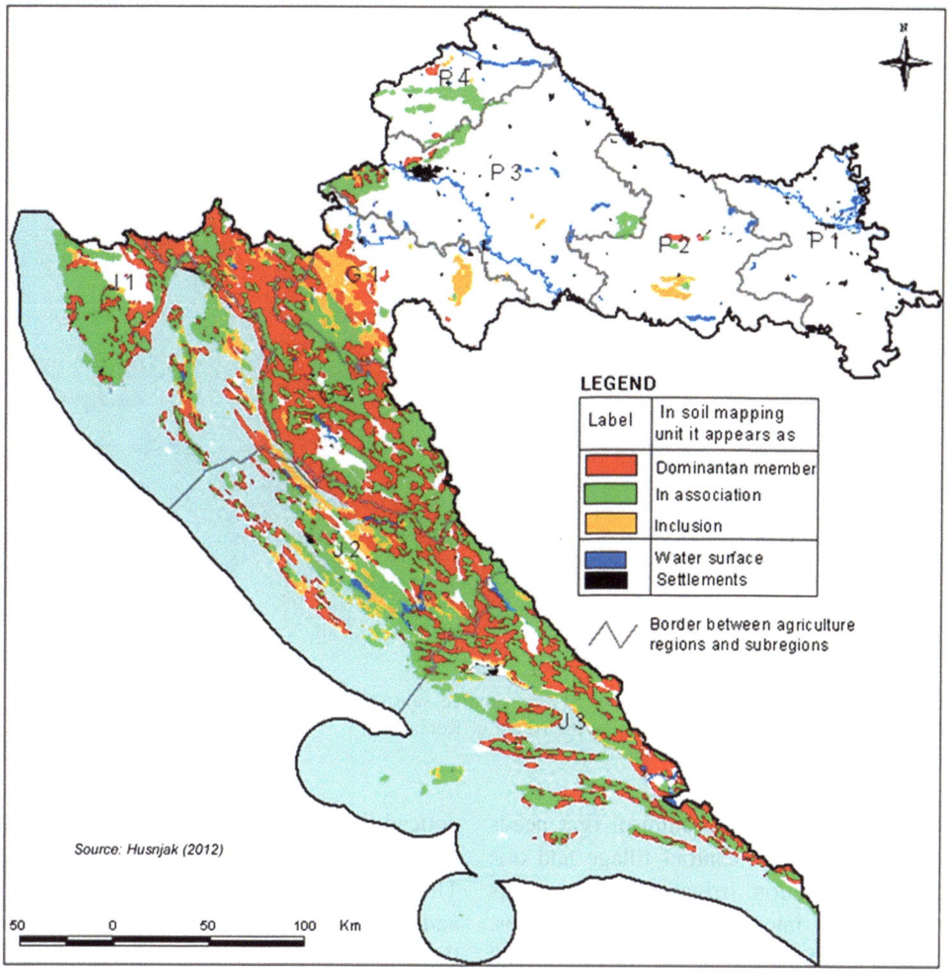

Fig. 6.14 Distribution of kalkocambisol in the pedosphere of Croatia

Table 6.14 Texture and chemical properties of kalkocambisol

Soil horizon	pH in H_2O			P_2O_5 mg/100 g of soil			K_2O mg/100 g of soil			Humus (%)			Clay content (%)		
	Average	Minimum	Maximum	Average	Minimum	Maximum	Average	Minimum	Maximum	Average	Minimum	Maximum	Average	Minimum	Maximum
A_p	5.4	3.9	7.2	1.4	0.0	8.2	17.5	6.2	45.0	9.6	5.9	15.8	36.7	10.5	56.2
$(B)_{no}$	6.3	3.6	6.6	0.6	0.0	2.5	9.2	3.7	18.0	4.4	1.5	9.7	48.5	24.4	80.4

Source Bašić et al. (2003)

is higher, but within a shallower horizon, at 5–10 %. Our observations, which of course should be checked against climate change, are that the extreme droughts and extremely high temperatures in particular, as well as the large number of forest fires, have caused irreversible coagulation of mineral colloids in deep horizons of these soils, which increases the stability of the structure and permeability to water. In this way they have increased permeability to water so there is reduced surface runoff and erosion, but higher leaching.

Land use. By far the largest area is used as pastures: Mediterranean maquis and forest garrigue of low value in the Adriatic region under the influence of a Mediterranean climate. In contrast, in the Mountain region there are excellent and very good pasture and forest habitats, whose value depends on altitude, inclination, exposition, and, mostly, on soil depth. They are used as arable land, but the yields are low and unstable, especially in dry years. With irrigation they are a very good garden soil. However, in karst areas and their peripheral parts, the vineyard soils are excellent (Photo 6.23).

Photo 6.23 Typical deep kalkocambisol, on the plain, a very suitable substrate for vines—Savudrija in Istria (J-1). *Right* The typical "Mediterranean polyculture" on kalkocambisol—Blato on the island of Korčula

Guidelines for management. Kalkocambisol first needs efficient protection from erosion—contour tillage and orientation of rows of the plantation. Irrigation of these soils opens the door to use in the intensive cultivation of vines, olives, figs, citrus fruits, and vegetables on open fields and indoors. Fertilization is especially important; it is recommended to use phosphoric fertilizers.

6.1.4.2 Red Soil (Terra Rossa)

The name of the soil type. The name comes from the color of the soil (Italian: *terra*—soil, *rossa*—red). This soil is also described as Mediterranean red soil, rhodochromic cambisol, and cambisol rhodic. When speaking of terra rossa we are speaking of soil of red color generated on limestone. Excluded are all soils formed on substrates of red color, which, except for the red color, have nothing in common with red soil formed on limestone.

Genesis and sequence of horizons. Red soil is a soil of A_{oh}–$(B)_{no}$–R sequence of horizons in profile, with a characteristic red color and heavier texture, which means high clay content and sharp-edged polyhedral structures. The structural units (aggregates) have a special flashy film on the surface. Soil mass is non calcareous in spite of forming on limestone. Terra rossa is generated on pure Cretaceous limestone. Its genesis and place in the evolutionary succession of red soil is one of the most studied issues in the field of soil genesis in Croatia. For the genesis of red soil there are many theories. Our mineralogists noticed the similarity between red, bauxite residue and the insoluble residue of chemical weathering of limestone. They believe that the red comes from the insoluble residue, and that it is of indigenous origin. Gračanin (1951) studied the genesis of terra rossa. He believes that terra rossa is an old and relict formation created by the soil type-generating process of rubification, but this material under recent conditions can maintain an unchangeable morphology and the same or similar properties, or is altered, so it becomes the parent substrate for soil genesis from which will develop new soil types. Gracanin describes the process of rubification as several separate processes: decarbonatization ($CaCO_3$ leaching), desilication (leaching of silicon similarly to lateritization), and the accumulation of iron and aluminum (siderolitization); the processes described taking place under the tropical climate conditions that prevailed until the Pliocene period of the Neogene—Tertiary. According to Gracanin, terra rossa is a relict of genesis under tropical conditions of the Paleocene—Tertiary. Under Mediterranean climate conditions terra rossa soils show high stability and retain their typical properties, while in the inland–continental area, depending on recent weather conditions, they may retain their properties if the climate is similar to the Mediterranean, and if not terra rossa has the role of parent material which leads to cambic processes (chemical disintegration and formation of a cambic horizon type $(B)_v$, followed by leaching and acidification, continuing with eluvial–illuvial migration). This

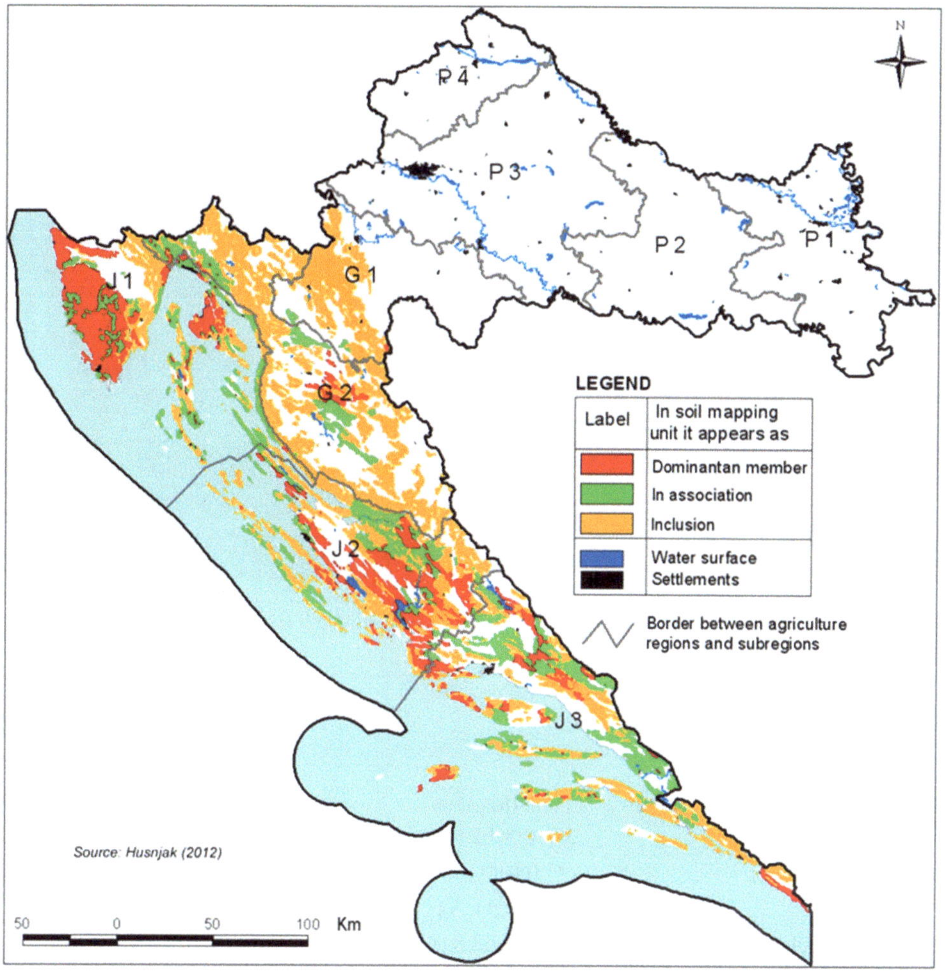

Fig. 6.15 Distribution of red soil (terra rossa) in the pedosphere of Croatia

is the case with the red soil of Lika and Kordun (G-1 and G-2 subregions).

On the basis of the results of this research all continental red soils on limestone or dolomite, which show signs of degradation (cambic process or eluvial–illuvial migration) are today usually called relict terra rossa. Gracanin systematized red soils as *typical* if formed by rubification, and *atypical* (lithochromatogenic) in the case of all other soils of red color formed on various substrates, or *zonal* (in the Mediterranean zone, without degradation) and *extrazonal*, which includes continental–relict red soil.

The WRB name. There is no adequate soil name; perhaps we may suggest "cambisol rhodic".

Position in an evolutionary series. There is no consensus on the evolution of red soil is unique. One view, which we accept, is that red soil is a member of the following evolutionary succession of red soil:

KALKOMELANOSOL (organic → organo-mineral → rhodic) → **TERRA ROSSA** (typical → luvic) → **LUVISOL**

Another opinion is that red soil as a parent substrate forms the evolutionary succession:

TERRA ROSSA (cambic → luvic) → **LUVISOL** (typical–acric) → **ACRISOL** (typical → pseudogleyic) → **PSEUDOGLEY**

Taxons below subtype are typical and luvic terra rossa. In a typical red soil there is no morphological sign of other pedogenetic processes, except for the characteristic type, and luvic shows a slight textural differentiation in its profile. It gradually creates an initial E horizon.

Distribution. In the Croatian pedosphere terra rossa accounts for 245,289 ha or 4.4 % of total land area. It occurs in the area around the Mediterranean (Greece, Italy, France, Spain, and Portugal), and in the coastal regions of Croatia, the Slovenian Littoral and Istria to Montenegro, and on the islands. The relict form we find in the continental part, the Mountainous region of Lika and Kordun (G-1 and G-2 subregions) (Fig. 6.15).

Terra rossa is the dominant member of mapping units in the largest part of Istria, the northern part of Krk island, in the Ravni Kotari near Zadar, and then in the continental hinterland of Split and the islands of Brac and Vis. Relict red soil occurs as the dominant unit west of Otocac in Lika, and as inclusions over the whole area of the lowlands of G-2 subregion, except in

Photo 6.24 Underground limestone relief leads to heterogeneous depth (*left*). *Right* Deep terra rossa with bauxite—Karojba near Motovun in Istria

Table 6.15 Taxonomy of terra rossa

Name of taxon	
Variety (based on depth of soil to unchanged limestone)	Form (based on topography slope in %)
<35 cm (shallow)	<3 (plain)
	3–8 (slight slope)
35–70 cm (medium deep)	8–16 (moderate slope)
	16–30 (moderately steep)
>70 cm (deep)	>30 (steep)

karst areas. The association occurs as inclusions in the region and across the Adriatic islands (Photo 6.24).

Physical and chemical properties. Physical and chemical properties depend on the depth of the red soil and its evolutionary stage. In the typical red soil, all the characteristics required for the soil to be used for growing crops in agriculture are very good. These features are reflected by the data in Tables 6.15 and 6.16).

Red soil contains large amounts of predominantly kaolin clay with a little vermiculite, but yet so high a clay content that, if properly managed, the stable structure of this soil is the basis of very favorable physical properties—water-holding capacity, infiltration of rain water, and thermal properties. The soil is warm, loose, and permeable to water. The stability of the structure is explained by the saturation of the CE colloidal complex by Ca^{++} and Mg^{++}, which are very efficient coagulators of microaggregates in the first stage. Cementation of microaggregates and the formation of macroaggregates occur by irreversible coagulation of iron and aluminum hydroxides in the second stage. Red soil has a favorable ratio of micro- and macropores, good permeability to rainwater, and aeration. The water-holding capacity is high. The chemistry of red soil is practically neutral to slightly acidic, having a high CEC (30–60 meq/100 g of soil); the degree of saturation (mainly Ca^{++} and Mg^{++}) is above 80 %. Subtype luvic red soil is acidic and the base saturation is significantly lower. Because of the low content of humus red soil is poorly supplied with nitrogen, but particularly poor in available phosphorus, because it binds to barely soluble iron and aluminum phosphates. The supply of potassium is moderate. Because of the very favorable physical, chemical, and biological characteristics

Table 6.16 Important physical and chemical features of red soil luvic

Soil horizon	pH in H_2O			P_2O_5 mg/100 g of soil			K_2O mg/100 g of soil			Humus (%)			Clay content (%)		
	Average	Minimum	Maximum	Average	Minimum	Maximum	Average	Minimum	Maximum	Average	Minimum	Maximum	Average	Minimum	Maximum
A	5.9	4.7	7.2	0.6	0.0	1.6	14.3	7.4	22.0	9.3	4.0	17.7	46.9	28.7	63.3
E	5.7	4.0	7.0	0.0	0.0	0.1	9.1	4.8	14.0	3.0	1.4	4.7	60.2	33.4	81.2
B_t	5.8	3.9	6.6	–	–	–	–	–	–	–	–	–	75.9	53.8	85.0

Source Bašić et al. (2003)

Photo 6.25 The special form of vine breeding of the Peršurić family near Poreč in the area known as "Red Istria" (*left*) and intensive olive growing—"Plava laguna"—Poreč on terra rossa (*right*)

and high fertility, red soil bears the attribution "Chernozem of the Mediterranean."

Land use. All Mediterranean cultures are successfully grown on red soil—olive, grape, fig, citrus, cherry, hazel tree, peach and nectarine, and all vegetable crops and tobacco. All grown fruit and crops are of exceptionally high quality. Olives grown in red soil provide superior quality of olive oil. Yield and its stability depend on soil depth. From an economic point of view, growing early varieties of grape, fruits, and vegetables is of particular interest. Grown on terra rossa, wine varieties, specially the famous and superb Istrian wines Teran and Malvasia, have a special unforgettable taste. Compared to other Istrian soils—regosol or rendzina on flysch and terra nera—wine on red soil is mild with lower acidity (Photo 6.25).

The development of tourism increases demands on the supply of quality products—wine, olive oil, fruits, and vegetables—which opens the door to possible exploitation of terra rossa as some kind of trademark, e.g., "Red Istria."

Guidelines for land management. In spite of the fact that the value of red soil in agriculture depends on the depth of this soil, it is necessary to take into account that the luvic subtype is less fertile than the typical one, because of leaching of bases and plant nutrients. The only permanent limitation in terms of cultivation of agricultural crops is the permanent lack of water, especially in the growing period when both the requirement for water and the consequence of its deficit are maximal. Therefore, investment in irrigation systems, like all other ameliorative practices on red soil, is economically justified. Without irrigation, management has to be predominantly water conservation-oriented, which means that agricultural activities need to ensure soil conditions that enable

Photo 6.26 "Partners in the fight for survival": With its crown and root system, pubescent oak (*Quercus pubescens*) protects "the island" of terra rossa, which "pays it back" by supplying water and nutrients as the basis of existence of the oak—Lim channel in Istria

infiltration of a maximal quantity of water from the autumn–winter period and protection against evaporation during the growing season. Despite these limiting factors, thanks to plenty of light and sunshine and the large number of sunny days, this soil has a high economic value.

Terra rossa is prone to erosion, and the greatest damage by erosion is from storms—torrential rainfalls, which are a regular occurrence; this erosion is greater if the soil is left without vegetation cover and exposed to torrential rains. Effective protection is achieved by vegetation cover, so it is advisable to manage this soil with a minimal period of no vegetation cover. The photos illustrate the importance of vegetation in the protection of red soil (Photo 6.26).

Natural features of red soil are the inherently insufficient supply of plant-available phosphorus, far below the needs of most crops. Therefore, fertilization with this nutrient has to be one of most important goals in management on this soil, which will respond to this practice. Humus content in the red soil used in agriculture is inadequate and the management of humus is a key goal in maintaining favorable conditions.

It is important to stress that one obligation of sustainable soil management is to avoid any traffic in the wet state and compaction of the red soil, which normally occurs as a necessity for protection of vines. The consequence is the long-term deterioration of physical properties.

6.1.5 Class V: Eluvial–Illuvial Soils of A–E–B_t–C/R Sequence of Soil Profile

Increasing amounts of rainfall and increased leaching of all cambic and polygenetic soils leads to debasification and acidification. Upon reduction of the content of bases in soil solution and in the colloidal adsorbing complex, the pH of soil gradually falls. When the pH of soil drops below pH 6, a base saturation of adsorbing complex below 70 %, this leads to peptization of colloidal clay particles. These particles peptizated in water migrate from the initial eluvial E horizon in a descending direction, and are retained in the deeper illuvial B horizon at a pH known as the isoelectric point in which, because of coagulation, the migration of colloids ends. The process is called eluvial–illuvial migration; the eluvial E horizon is light gray and impoverished, and the illuvial B_t horizon is enriched with clay. Thus, there is a gradual textural differentiation, because the entire migration of particles of clay occurs without destruction. In reviewing the previous class of soils we have seen that this process starts at the stage of cambic soils. The process is similar to podzolization with eluvial–illuvial migration of clay particles, but that occurs after prior chemical decomposition of the clay, so that the products of decomposition, i.e., the components of clay minerals—sesquioxide R_2O_3 (Latin *sesqui*—one and a half, which means that for every atom of Fe or Al there is one and a half atoms of oxygen)—migrate separately. Moreover, this leads to migration of organic colloids—humus. The endomorphological consequence of this process is the formation of a marked spodic (*spodos*—ash) E horizon, with a lighter (ash-gray) color, rich in SiO_2, which is insoluble under the same conditions and is not the subject of descedent migration. Furthermore, we have an illuvial B horizon, which is labeled B_h if humus accumulates in it, or B_{fe} if sesquioxide accumulates. Analytically it is simple to identify the process of podzolization because the ratio SiO_2:R_2O_3 in the eluvial horizon is in favor of SiO_2, and a higher content of R_2O_3 occurs in the illuvial horizon of podzols. Luvic and podzolic processes mutually differ therefore in terms of the mechanism or type of substances that migrate. Until recently the boundary between these two processes was not sharp, and they were indicated as luvic or podzolic soils.

6.1.5.1 Luvisol

The name of the soil type. The name is derived from the word *luvic*—flushed or washed out, which describes the main process of clay migration resulting in textural differentiation of the profile. It is therefore the impact on soils exposed to large amounts of rainfall. In our literature, these soils were initially described under the names podzolic brown soil, weak podzols, podzolic soil, and luvic brown soil. At that time no differentiation was made between luvic and podzolic processes; for both, the term podzolic was used. But there was also the name "real podzols." Later, in the former Yugoslavia state we accepted the French name *lessive* and described this soil type as lessive soil as distinguished from podzols.

Genesis and sequence of horizons. We have an E horizon (from *eluvial*) and brownish colored illuvial B_t horizon named argiluvic (*argila*—clay, *luvic*—washed out). Compared to the eluvial E horizon, it has higher clay content and a characteristic thin, glossy-reflective film around the surface of the structural aggregates formed by illuviated clay particles. It occurs under conditions of a somewhat wetter climate than that of a cambisol eutric, i.e., more than 700 mm annual rainfall and a mean annual temperature of 8–11 °C. Outside this region it occurs only if the relief (topography) makes changes to the climatic conditions in the direction of higher wetting. It is most common in valleys and mild slopes up to 800 m above sea level, on limestone and other substrates.

This soil occurs on different parent substrates, especially loose, deep, light, non-calcerous loam. Rarely it is formed on heavier textured clay material, because of low permeability and sandy substrates, and because they have no clay as input "material" to eluvial–illuvial migration. For the formation of luvisol, a gradual leaching of bases is necessary in the previous stage of evolution of the soil. The basic processes of formation are as follows:

- We have the accumulation of humus and mineral elements in the surface horizon and the formation of a humus horizon of ochric—A_{oh} or umbric—A_{um} type.
- Eluvial–illuvial migration of clay particles. This process starts in the stage of soil evolution after leaching and washing out of ion coagulators of colloids (Ca^{++} and Mg^{++}) and acidification. The peptization of clay, which is a precondition for eluviations, starts in the pH range 4.5–6.5.
- Suspended in soil solution, clay minerals move in peptizated form through macropores in a descending direction in which the pH gradually increases. At the illuvial B_t horizon, when the pH reaches the isoelectric point, clay particles coagulate forming a film on the surface of the structural aggregates, visible like a glossy coat.

At conditions below pH 4.5, Al^{+++} and Fe^{+++} ions are released, which perform coagulation and prevent leaching. Each soil, during natural evolution and the transition from young to mature stages has entered, this stage of soil evolution. Thus, the longer the soil pH is retained in this range, the more clay is moved by this mechanism, leading to impermeability of the illuvial B_t horizon.

A sudden decline in the concentration of ions in soil solution after heavy rains or melting snow causes a decrease in the stability of the clay particles—peptization and eluviation in a descending direction. Low-molecular-weight organic compounds formed in the process of humification, especially under low pH (acidic) conditions, are capable of binding with aluminum and iron in the soluble compounds. Their leaching reduces the stability of clay—leading to peptization and eluviation. Mechanical movement of clay through torrential rain is possible in the dry summer months. The sudden and heavy rain carries the particles of clay that fill the pores of the soil and deposits them in the illuvial B_t horizon, gradually making it impermeable to rainwater.

Thus, the process of downwards movement of clay occurs to a greater or lesser extent in all soils, but not always by the mechanism of eluvial–illuvial migration. In luvisol it is the dominant process, resulting in endomorphological and textural differences in the soil profile: an E horizon of light-gray color and lower clay content, and a B_t horizon of yellow-brownish color, enriched by clay particles.

Table 6.17 Taxonomy of luvisol

Name of taxon	
Variety (based on process of genesis)	Form (based on texture of soil)
For all subtypes • Typical • Pseudogleyic • Podzolic	For all varieties • Loamy • Sandy • Clayey

Generally, the migration of smaller clay minerals, such as montmorillonite, is faster, followed by illite and kaolin. It is estimated that, depending on the parent substrate, 5,000–10,000 years is necessary to form luvisol. All Croatian luvisols occurred after the last glaciations; cambisol eutric is younger, but chernozem and leptosol on loess are younger than cambisol. Regosol is practically a recent soil.

The WRB name. Luvisol

Position in an evolutionary series. The evolution of luvisol takes place in two directions:

- On lighter, loose, loamy, calcareous parent substrates, clay accumulation in the B_t horizon at a certain stage causes low permeability to rainwater, leading to initially short-term but then long-term stagnation and pseudogleyic processes. Its evolution is therefore in the direction of pseudogley formation. On such substrates as loess and loess-like sediments, we have the following evolutionary series:

CAMBISOL EUTRIC (typical → luvic) → **LUVISOL** (typical → pseudogleyic = Luvic pseudogley) → **PSEUDOGLEY**

- On lighter, porous, and acidic substrates evolution takes place in the direction of podzols, with the series:

CAMBISOL DYSTRIC (typical → podzolic) → **PODZOL**

The taxonomy of luvisol at units below subtype is presented in Table 6.17.

Luvisol typical is a soil without the morphological characteristics of any other processes except those typical for this type of soil. Typical luvisol on limestone has A and E horizons, whose depth does not exceed 1/4 of the total depth of the soil profile.

Luvisol pseudogleyic is a transitional stage of evolution between luvisol and pseudogley on siliceous–carbonate substrates, such as loess and Pleistocene loam or metamorphic parent substrates rich in clay silicates (some shales and phyllites). A characteristic of this variety of luvisol is occasional, mostly short-term retention of rainwater in the E horizon. This phenomenon manifests itself morphologically in the form of the rare occurrence of soft dark brown concretions of manganese and humate and bluish-gray and rusty greasing as a consequence of the alternation between

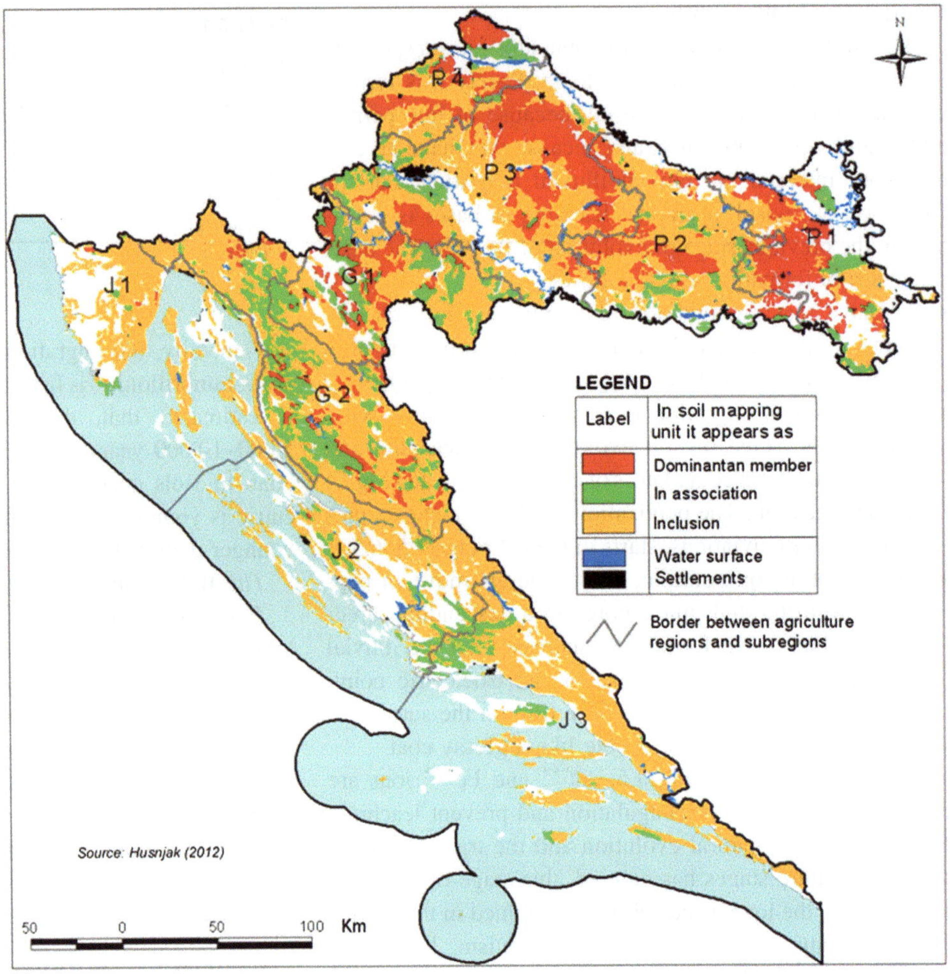

Fig. 6.16 Distribution of luvisol in the pedosphere of Croatia

maximal saturation by rainwater in the wet stage (reduction) and aeration (oxidation) in the dry stage. Because of the low permeability, there is no descending but instead only lateral movement of reduced compounds.

Luvisol podzolic is mostly forest soil with appearance of organic (O_l and O_f type) subhorizons, followed by an A and spodic horizon, morphologically recognizable by particles of quartz sand of bright color as a sign of the genesis of an initial spodic E horizon.

Distribution. With an area of 703,215 ha or 12.6 % of the territory, luvisol is the most abundant soil type of the Croatian pedosphere. On this basis it can be concluded that bioclimatic conditions in the major part of Croatia correspond to the climax stage of pedogenesis of luvisol. In the Pannonian region it occurs as inclusions in the P-1 subregion of the wavy eolian-shaped relief of the Vukovar loess plateau. It is a natural cambisol eutric soil zone, but to the west of the P-1 subregion, on border areas with annual rainfall higher than 700 mm, it has already emerged as the dominant unit on the leached Pleistocene loess and loam (Fig. 6.16).

By inspection of the figure we learn that luvisol is the only soil type that occurs in all agricultural regions and subregions of Croatia, and is the dominant unit in the cartographic unit in all subregions of the Pannonian region. The association has a significant presence in the mountain regions, especially in subregion G-2, which represents a valuable substrate for the extremely valuable mountain forests (Photo 6.27).

Physical and chemical properties. Because luvisol represents the climax stage of pedoevolution, it is formed on practically all parent substrates of Croatia, which means there is maximal heterogeneity of all properties of this soil type. They differ based on the different substrates, topographic position, exposition and inclination, moisture, vegetation, and form of use. The silicate substrates generally have less clay, and textural differentiation is less pronounced, in contrast to the siliceous limestone, which is very pronounced. The same goes for luvisol on limestone and dolomite, which shows a strong differentiation of structure, but the illuvial horizon still retains some water

Photo 6.27 *Left* Luvisol typical on loess—Županja (P-1), *right* luvisol pseudogleyic on Pleistocene loam—Grubisno Polje in the P-3 subregion

Table 6.18 Some physical and chemical properties of luvisol on loess of the Pannonian region

Soil horizon	pH (nKCl)			mg P_2O_5/100 g of soil			mg K_2O/100 g of soil			Humus (%)			Clay (%)		
	Average	Minimum	Maximum	Average	Minimum	Maximum	Average	Minimum	Maximum	Average	Minimum	Maximum	Average	Minimum	Maximum
A	4.5	3.5	5.8	6.3	0.5	20.5	14.2	9.0	23.5	13.9	9.2	19.6	24.6	10.2	27.2
E	4.3	3.4	5.4	4.2	0.5	15.0	6.8	2.2	11.0	4.0	1.6	6.4	21.6	12.2	28.1
B_t	4.3	3.8	5.1	–	–	–	–	–	–	–	–	–	27.8	20.4	35.0

Source Bašić et al. (2003)

Table 6.19 Some physical and chemical properties of luvisol on limestone and dolomite of the G-2 subregion

Soil horizon	pH (nKCl)			mg P_2O_5/100 g of soil			mg K_2O/100 g of soil			Humus (%)			Clay (%)		
	Average	Minimum	Maximum	Average	Minimum	Maximum	Average	Minimum	Maximum	Average	Minimum	Maximum	Average	Minimum	Maximum
A	5.1	4.3	6.8	1.9	0.0	7.0	22.2	13.5	40	11.5	3.0	37.1	23.2	9.4	46.9
E	4.4	3.7	6.4	0.2	0.0	1.2	8.4	3	15.7	2.6	0.9	7.7	31.0	17.2	69.9
B_t	4.6	3.8	6.1	–	–	–	–	–	–	–	–	–	48.5	25.2	82.3

Source Bašić et al. (2003)

permeability. Therefore, the evolution of luvisol on limestone is not directed towards forming pseudogley, except as a relict soil. For these reasons, separate tables present the characteristics of luvisol on loess (Table 6.18) and luvisol on limestone and dolomite (Table 6.19).

Textural differentiation in luvisol on loess is obvious. It has adverse physical features, which are more unfavorable due to the high silt content, which causes deterioration of the structure, a tendency for crust formation, and high soil erodibility.

The result, of decarbonatization, leaching, and acidification are unfavorable chemical properties of luvisol, such as acidic or very acidic soil chemistry, high potential acidity, low CEC, and low base saturation.

If cultivated, this soil has a low content of humus and insufficient supply of plant-available nutrients. The differences in characteristics between luvisol on dolomite in relation to that on loess and other substrates are not small. First, luvisol on limestone contains more clay minerals throughout its profile, and the textural differentiation is apparent. It is

Photo 6.28 Rainfall of high intensity causes the crust formation of luvisol pseudogleyic; before planting (*left*) and after crop emergence (*right*)—Garešnica in the P-3 subregion. Breaking the crust using the row cultivation procedure is required

apparent that, the clay content in the E horizon is still quite high. Here, however, kaolin dominates, while illite dominates on loess. However, the maximum content of clay in the illuvial horizons of 82 % is very high (Photo 6.28).

Luvisol on dolomite has a heavier soil texture than that on limestone. We did not find an explanation for this difference, but are sure that genesis and some specific properties of soils on pure dolomite need more research. Dolomite as a parent substrate is a little marginalized from the taxonomical point of view, where the definition of kalkocambisol includes cambic soils on dolomites. In addition, dolomites in Croatia are not homogeneous rocks. Coming back to luvisol, we would like to point out that in the dry state, the illuvial horizon of luvisol on dolomite breaks into larger clumps of polyhedral form. Because of the texture, the pH value is slightly different, less acidic, and it contains less plant-available phosphorus, which is a general characteristic of soils on limestone and dolomite. However, it contains an abundance of available potassium. The humus content in the soil should be viewed in the light of the fact that in natural soil under natural vegetation the humus is concentrated in a shallow (just 4–5 cm) surface horizon, and on cultivated soil upon plowing is mixed into a much deeper layer, and a smaller percentage of content does not mean a lower total content in the soil.

Land use. Luvisol is used as a natural habitat for Croatian forests of the highest quality. The best mountainous forests of beech (*Fagus sylvatica*) occur only on this soil type. In agriculture luvisol is used for a wide variety of permanent plantation, as well as arable land (cereals, sunflower, sugar beet, and alfalfa), pastures, and meadows. The largest fruit plantation Borinci near Vinkovci and vines (Đakovo) are located only on luvisol on loess, of course after preparation of the substrate for plantation by liming, humization, and fertilization.

Guidelines for management. Compared with all other "younger" soils of this evolutionary series, luvisol is a soil with more negative features for crop growing: textural differentiation and limited rooting, the rhizosphere is formed in horizons above the illuvial B_t horizon, it has an unstable structure, favors crust formation, and sometimes heavy rain is followed by rainwater stagnation, acid chemistry, poor nutrient supply, etc.

The relatively low crop production value of luvisol as natural soil can be increased to the level of very good agricultural soil without significantly high investment. Ameliorative interventions are required for this purpose: vertically deep loosening or subsoiling, chiseling, conservation tillage, and a melioration of chemical features by liming, ameliorative fertilization, and humization. Liming is certainly a key practice, which opens the way for forming an affordable crumb structure and adequate water–air relationship in the rhizosphere. Generally, in the luvisol zone of the Pannonian region, the amounts of precipitation meet water requirements of agricultural crops, but on the eastern edge of this zone drought is not rare and irrigation is justified, especially for sensitive crops, for example vegetable crops. One should emphasize that the whole area is supplied by sufficient quantities of available quality water for the purposes of irrigation. Therefore, remedial measures are directed towards increasing the depth of the plow layer using deep loosening and plowing, repairing structure by humification, liming, fertilization, cultivation of green clover–grass mixtures in the crop rotation, and fertilization with high doses of mineral fertilizers. Inclined soils need protection against erosion. We recommend soil tillage along contour lines. Deep plowing mixes the soil material of the A, E, and part of the B_t horizon and creates a much deeper physiologically active profile with favorable water–air conditions. After the performance of amelioration, repairing the physical, chemical, and nutritional characteristics using mineral fertilizers, it may be possible to achieve stable and high yields of all crops on this soil. As part of the management of this soil one should note the permanent principle that every time the soil passes into an unfavorable moisture state, i.e., the soil is too wet, this always causes more harm than good, and therefore this should be avoided. A key management factor is a high content of humus and microbial activity, which is why there should be adequate coverage of livestock units per hectare and standard use of farmyard manure. A favorable circumstance is that these soils are so pure that one can practice and develop all forms of ecological agriculture. As both objectives cannot be achieved in a short period, transition to a recommended system of management in this area takes a relatively long time.

6.1.5.2 Podzol

The name of the soil type. The name is of Russian origin (*zola*—ash), related to the soil's ashy color. In Croatian literature several soil types were described under this name: luvisol, pseudogley, and podzols as we define them today. The concept of podzols is today narrowly defined and refers to soil with substantially different processes of genesis and properties compared with luvisol and pseudogley.

Genesis and sequence of horizons. The soil profile of podzol forms O_l and subhorizons, followed by an umbric humus-accumulative A_{um} horizon containing acidic humus in which fulvo acids dominate. It has a black color if wet but is ashy or gray if dry. Then follows the most characteristic horizon of podzols—the spodic E horizon of ashy color with characteristic and significant quartz sand particles. Quartz sand occurs after the destruction of the mineral complex of the soil. Podzols form only on poor, acidic, and highly porous silicate substrates under cold and humid climate conditions with rainfall exceeding 1,000 mm per year. These conditions are generally preferred by species of coniferous forest vegetation whose residues are poor in bases. This favors the formation of crude humus with a high content of low-molecular-weight fulvic acids and strong processes of destruction of the mineral component of the soil. The illuvial horizon may contain separate subhorizons of illuviation of humus—B_h and iron—B_{fe}, or only B_{fe}. The soil is very acidic (pH in water is below 4), and it has a low CEC and degree of base saturation.

Table 6.20 Taxonomy of podzol

Name of taxon and base of taxonomy	
Variety (based on character of illuvial B horizon)	Form (based on depth of spodic E horizon)
Ferrous	Weak (<10 cm)
Humus–ferrous	Moderate (10–20 cm)
	Strong (>30 cm)

Position in an evolutionary series. Podzol is the last, i.e., the oldest member, of an evolutionary series of soils on acid rocks (Table 6.20):

RANKER (typical → cambic) → **CAMBISOL DYSTRIC** (typical → podzolized) → **PODZOL**

WRB. Podzol

Distribution. Although it is highly prevalent in the world (northern Europe, Scandinavia, North America), in the Croatian pedosphere podzol occupies only 1,382 ha, or only 0.025 % area in Gorski Kotar and at the top of Moslavacka hill (Fig. 6.17).

At two locations in Gorski Kotar it occurs as the dominant member, and in association with other members of the evolutionary series of soils on silicates in other areas. Over a negligible area we found it on the metamorphics of Kalnik (Photo 6.29).

Physical and chemical properties. Podzol is a soil of light texture in which there are fractions of coarse and fine sand with the chemical composition of silicon dioxide (quartz); it is very permeable to water and the soil is well aerated. The soil chemistry is very acidic, with a pH ranging from 3.2–3.7 KCl, it has a low CEC and base saturation, the soil is leached and poor in nutrients. Somewhat better is the supply of plant-available potassium, but it is poor in phosphorus, since phosphorus is a component of the insoluble mineral compounds which disappear under very acid conditions. Humus of the shallow A_{um} horizon reaches 50 %, dominated by raw humus.

Land use. Podzol is a typical forest soil; where it occurs, it has representative acidophilic vegetation, in the first place the well-known delicious forest fruits *Vaccinum myrtilus* and *Fragaria vesca.* In our country it is used as a forest soil, and to a lesser extent for pasture and arable land, on which are grown barley, oat, and potatoes.

We think of this soil type more as a natural phenomenon, the last stage of pedoevolution, than as a valuable resource.

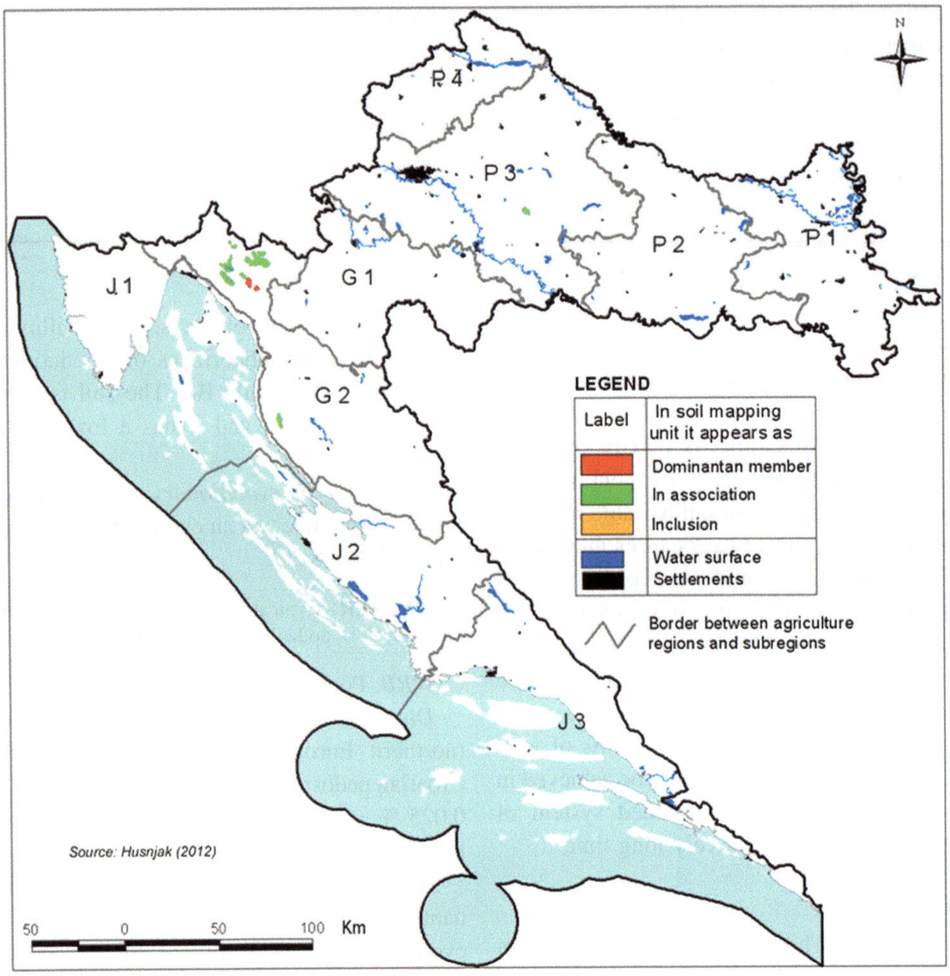

Fig. 6.17 Distribution of podzol in the pedosphere of Croatia

Photo 6.29 *Left* Humus iron, and *right* iron podzol in Gorski Kotar (Photo: Pernar). Because of soil movement it has a generally characteristic uneven boundary between the E and B horizons

6.1.5.3 Acrisol

The name of the soil type. The name is derived from *acric*—acid, which means leached and acidified soil. In our literature it is described as a double layer soil—luvisol subtype—luvisol acric, also called heathery or heathery-ferny soil (Stritar 1966), based on the natural vegetation that occurs on this soil. Only in the recent classification of Husnjak (2012) was this soil set aside as a special type.

Genesis and sequence of horizons. Acrisol is formed on the already-described layer called loess-like mineral material, with uneven thickness which covers relict soils in the Mountain region. It is believed that the loess-like material is eolian loess, formed in the late Pleistocene and early Holocene. Because of the high quantity of rainfall and high permeability, all bases are leached, washed out from it, so the base saturation in the colloidal CE complex is much lower than 50 % and mostly below 35 %, and lies on the B horizon whose texture with more clay is not the consequence of eluvial–illuvial migration, but is originally metamorphosed relict terra rossa. It occurs exclusively in the Mountainous region (both subregions) in Kordun, Lika, and Gorski Kotar, and in Istria, where this soil is the habitat of the well-known sweet chestnut—Istrian maroon. The sequence of horizons is O–A_{oh}–E–B–C, in which the B horizon is relict terra rossa which is sometimes pseudogleyic, meaning that it is impermeable to rainwater, which is the main difference between it and "real" terra rossa of the Adriatic region—Mediterranean.

Position in an evolutionary series. There are still questions that remain unanswered in relation to the genesis and evolution of acrisol. The loess-like materials show differences in terms of depth, but much more in terms of on which material it was deposited. According to our observations, the material falls either on older Pleistocene loam, clay changed by glaciations and solifluction, or it falls on the relict terra rossa which was also altered in the Pleistocene. Today some of these materials are permeable, others less so, disregarding them not retaining rainwater under recent conditions. Taxons below the level of soil type include the subtypes *typical*, which means without any sign of glacial processes, and *pseudogleyic*, which means endomorphological signs of pseudogley. Taxons below subtype are defined as shown in Table 6.21.

Of course, there are significant and clear differences between acrisol inside and outside sinkholes, especially in relation to the depths of loess-like and relict soil material. It is much deeper in sinkholes and generally negative forms of relief (depressions, karst fields) compared with outside sinkholes, but the inflow of rainwater in karst sinkholes is significantly greater and leaching is more intense.

Table 6.21 Taxonomy of acrisol

Name of taxon and base of taxonomy	
Variety (based on depth of loess-like layer)	Form (based on topography)
For all subtypes • 30–45 cm (shallow) • 45–60 cm (medium deep) • 60–70 cm (deep)	For all varieties • In karst sinkholes • Outside karst sinkholes

The WRB name. Luvisol acric

Distribution. Its distribution is very small in associations or as inclusions exclusively in the Mountainous region and the Istrian A-1 subregion as shown in Fig. 6.18.

At both sites it occurs in association with soils on limestone, dolomite or proluvial moraines and sediments. As a rule, acrisol is maintained in positions protected from erosion.

Physical and chemical properties. The physical and chemical properties are shown in Table 6.21.

The chemical features are unfavorable. Firstly, the pH value shows a strongly acid chemistry, the soil is poor in all nutrients, and because of the extreme acidity the A_{oh} horizon contains a lot of raw acidic humus formed by residues of the vegetation mass of fern (*Pteridium aquilinum*) and heather (*Caluna vulgaris*). All these negative features can be easily and inexpensively changed in the desired direction (Photo 6.30, Table 6.22).

Land use. Because of its depth, this soil is naturally very favorable for forest vegetation. This vegetation was in the past, under previous economic–social and historical conditions, cleared for arable land, and later abandoned and transformed into extensive public pasture of low economic value, although there was potential for forestry and arable crops (rye, wheat, and maize of some groups), and vegetables (potato, cabbage, and carrot).

Guidelines for management. Acrisol is an acid soil of high potential fertility. Its physical properties are favorable, but for use in agriculture soil reclamation with primary attention paid to the chemical complex of the soil should be applied. Liming has to be followed by use of phosphorus fertilizers, farmyard manure, growing of clover–grass mixtures (in the livestock areas), and green manuring with crops of high biomass (e.g., rauola-oil rap) with working of crop residues into the soil. According to the results of our research, acrisol requires large amounts of poorly soluble phosphorus fertilizers, combined with abundant application of organic fertilizers (mainly solid manure), in order to hinder or reduce harmful phosphorus fixation. After implementing these procedures and performing bacterization, we managed to create a favorable substrate for

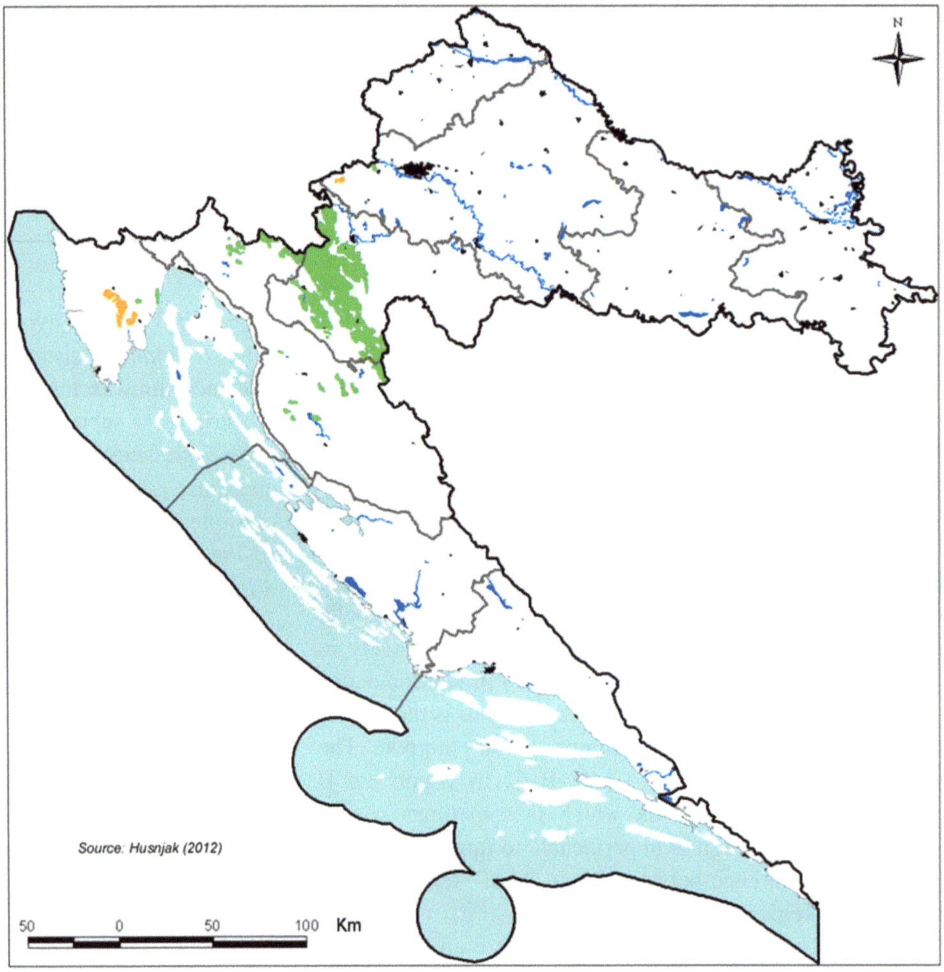

Fig. 6.18 Acrisol (two layers) in the pedosphere of Croatia

Photo 6.30 Acrisol typical on relict Terra rossa (*left*). Specific heathery vegetation on acrisol—Zir along the Zagreb—Split highway

growing alfalfa (*Medicago sativa*), a highly sensitive species (Butorac et al. 1988). As part of the same work we also noted that ordinary tillage with liming in a short time very efficiently removes the fern (*P. aquilinum*).

In spite of the potentially very high fertility of acrisol, it is not exploited in agriculture. Acrisol is an extremely favorable soil for some forest varieties, especially for conifers including "rapidly growing" varieties such as douglasia (*Pseudotsuga douglasi*) (Mayer 1984, 1992).

6.1.6 Class VI: Autochthonous Terrestrial Anthrosols of P–C, P–(B)–C, or P–B–C Sequence of Soil Horizons

The class of autochthonous terrestrial anthrosol is a new class in Croatian soil taxonomy. This class includes cultivated soil types formed in situ, and all changes in it are a consequence of the treatment of the soil for growing annual and perennial or permanent crops. In this class of Croatian

Table 6.22 Some indicative physical and chemical properties of acrisol

Soil horizon	pH (nKCl)			mg P_2O_5/100 g of soil			mg K_2O/100 g of soil			Humus (%)			Clay (%)		
	Average	Minimum	Maximum	Average	Minimum	Maximum	Average	Minimum	Maximum	Average	Minimum	Maximum	Average	Minimum	Maximum
A	4.1	3.9	5.2	1.4	0.0	5.3	17.7	5.6	53.4	7.4	4.2	12.7	25.2	11.1	44.4
E	4.4	3.6	5.7	0.5	0.0	2.3	6.7	2.2	24.0	2.3	0.5	6.0	29.5	19.8	44.7
B_t	4.9	3.7	6.6	0.7	0.4	0.9	10.3	9.0	12.3	2.3	0.7	5.4	43.5	23.4	66.7

Source Bašić et al. (2003)

soil taxonomy, Husnjak (2012) classified soils where soil material of more than one genetic horizon is mixed through tillage, creating on the topsoil an anthropogenic P horizon.

The anthroposphere is the inhabited part of the land where man is absolutely the central unit, the most powerful and influential biological factor of pedogenesis. The anthroposphere began with the beginnings of agriculture and man's sedentary lifestyle. Until then, man was part of nature. With agriculture and arable land began the removal of some natural plants which were declared undesirable and harmful, and the reinforcement of others—i.e., they were cultivated as desirable. The general opinion to date is that it all began in Mesopotamia, between the rivers Euphrates and Tigris and the river Nile, then in India, China, etc. Some archaeologists have questioned this and proved that the civilization of Mesopotamia lagged somewhat behind the civilization along the Danube including Croatian territory. Specifically, Jurić (2002, 2011) argues that between 5,000 and 4,500 years ago on the right bank of the Danube, in eastern Croatia, the area from Vukovar to Ilok (P-1 subregion) was occupied by indigenous populations, described first as the Vučedol culture (in relation to Vučedol locality near Vukovar), then the Vučedol civilization. The Vučedol civilization is a "European response" to the first "city-states" as a civilization leap, starting 5,000 years ago (Photo 6.31).

This culture was contemporary with Mesopotamia and the Old Kingdom of the Pyramids and the Egyptian empire. Jurić et al. (2000) argue for the origin of the Pythagorean theorem with this civilization, and this knowledge served for the building of Stonehenge and was transferred to Mesopotamia and Egypt. Unlike Sumer, tillage and cultivation of plants achieved cattle breeding on pastures of steppe. This culture produced the world's first bronze, and sought out copper ore. From the parent Vučedol civilization the population spread from Syrmium to the surrounding areas of Prague and the southern Carpathians in the Czech Republic to the north, the Romanian Banat in the east, the Alps in the west, and south to southern Bosnia. The first stellar calendar in Europe belongs to the same calendar age and the Egyptian and Mesopotamian calendar is called the Vučedol Orion. For the Vučedol people the disappearance of Orion from heaven meant the end of winter and the beginning of the year. All of these facts are presented to substantiate the claim that the fertility of the soils of the Vukovar loess plateau, similar to the fertility of soils along the Nile, drew the first humans, and the soil of this area has therefore been cultivated for over 5,000 years!

Man pushes the limits of his influence and affects soil in a myriad of ways via soil excavation and translocation during urbanization and infrastructure construction. However, industrial waste and other waste creates a substrate for the genesis of new soil, and these factors directly (irrigation, drainage) or indirectly (greenhouse gases) influence climate, remove natural vegetation, reduce biodiversity, and completely change the quantity, quality and the composition of the organic matter that will arise from the mold. The

Photo 6.31 The Vučedol dove as a "trademark" of the Vučedol civilization and a terrine with a calendar of the Vučedol civilization

Photo 6.32 In our coastal zone in the anthroscape on the island of Brac (Stipancici), man leaves his mark. It has changed and man's needs subordinated everything that Nature has created in past geological epochs. We are waiting the response of Nature! No doubt the response will come, but its form and quality is unpredictable

effects of anthropogenic change of relief can be radical. Through terracing on a more or less steep slope, man creates a flat surface, changing the landscape. This "scape" is no longer a landscape, instead a better term is anthroscape (Photo 6.32).

Assessing the anthropogenic impact on the genesis of soil, Resulovic et al. (2008) suggest that some impacts on the soil, the pedosphere, can be positive, while others can be retrograde or negative, in which case we talk about land degradation caused by mankind. Examples of retrogradation include soil drainage, stabilization of sandy soils, or moving sand, and other ameliorative practices which repair some properties of soil, returning soil genesis to "younger" stages of evolution and functions including non-productive ones. Conversely, soil degradation is the deterioration of some soil properties and its functions, like the use of practices which leads to a declining content and quality of humus, narrow crop rotation or monoculture, etc. Reduced humus content weakens the function of soil as a natural filter of water. As another example, plowing at the same depth compacts the layer below the depth of the plow, forming a plow pan. The short-term result on such arable soils is stagnating rainwater, but if the practice continues it is possible that an Antraquic horizon (WRB 2006) may form, which means an impermeable horizon without a luvic stage. We registered this problem in the practice of sugar beet growing with a narrow crop rotation. As a crop which needs more passes with heavy machinery for all crop growing practices, the growing of sugar beet means compaction of soil in a water-saturated state, especially in rainy autumns. Water stagnation on a compacted layer is a precondition for pseudogleyization, which leads to visible marks among step-by-step. It is usual that anthropogenic influence in the systematization of soils is under-evaluated. Therefore, although anthropogenic influence is profound in the pedosphere, in practically all soil taxonomies this influence is underestimated.

Soil is a natural body, and all changes from its natural state have to be properly registered in soil science so as to clearly address the origin and the scientific evaluation of the effects. As we described previously, one of the non-food functions is retention of different pollutants emitted from industry, traffic, urban structures, etc. Falling on the soil, pollutants move in the soil solution or suspension through macropores in a descending direction, but the next plowing, especially if it is deeper, will bring these pollutants up to the top surface again. This may be desirable from the point of view of water protection, but accumulation in soil and biological fixation in crops means it will become part of the food chain.

6.1.6.1 Rigosol

The name of the soil type. The name comes from the German word *rigol*—trenching, which means that this soil is formed by very deep plowing of a natural soil type.

Genesis and sequence of soil horizons. The main process of genesis of this soil type is formation of P (from English—*plow*) via deep tillage plowing and/or subsoiling or deep loosening, which mixes a minimum of two genetic horizons of the original soil, in combination with other measures like liming, fertilization, and cleaning of the skeleton. In the composition of the profile, one sees the characteristic P horizon, significantly changed in relation to the horizon on which it lies, and the horizon can be part of a deep humus accumulative horizon. They can lie on shallow parent substrates from which stone was removed and the P

Photo 6.33 Plow soil from shallow and stony kalkocambisol. Stones from the soil are removed by hand and built into "dry walls." This is really hard and sweaty work in the "desert"of karst limestone of the Adriatic hinterland

horizon is less than the surrounding skeletal soil, a skeleton that, when taken from the soil, is built into the dry walls.

Deep plowing and mixing affects multiple genetic horizons, and thus completely changes the natural soil; anthropogenic P creates a new horizon. It is of darker color, and differs from the horizon or substrate on which it lies; with this there is a sharp boundary or transition. Many of the natural features and morphological characteristics have been lost and they become new anthropogenically created features. The sequence of horizons is: P followed by C, a cambic or illuvial horizon, and C/R. On luvisol the P horizon contains A, E, and part of the B_t horizon. There are of course other combinations and possibilities. Identification—the origin of the material components from which P horizon originated—is not always easy to determine (Photo 6.33).

All features of these soils depend, of course, on the soil type from which they are generated and the characteristics of the horizons from which the P horizon is created. The only feature that is common to these soils is a higher content of plant-available nutrients and a pH value in relation to the natural soil type.

The WRB name. Anthrosol

Position in an evolutionary series. This depends on the one hand on the origin of the anthropogenized soil type, and on the other, on the climax stage of the natural soil in the area where the plowing is done. In the luvisol zone, evolution is driven in the direction of leaching, acidification, eluvial–illuvial migration, and textural differentiation, and in the chernozem zone in the direction of cambic soil. Criteria for taxonomy on the level of varieties and forms are shown in Table 6.23.

Distribution. Rigosol and hortisol occupy 217,370 ha or 3.9 % of the Croatian land area.

It is widespread in all regions and subregions of Croatia, as the dominant unit in all the vineyards of Ilok and Baranja in P-1, then in the Požega vineyards and vineyards in the foothills of Bilogora to Medjimurje and from Istria, through the islands to Konavle. Similar is the distribution in mapping units, where the association occurs as inclusions.

Physical and chemical properties. These are highly variable and depend on the natural features of the underlying soil on one hand, and the plantation management practices of these soils on the other. In relation to nature, the sources of the soils certainly show a difference in higher nutrient content.

Land use. This soil is used almost solely for permanent crops, but some arable crops like sugar beet need anthropogenization. We see deepening of the physiologically active profile of the soil; on some soil types changes are so radical that all properties of the original soil type change. By studying the depth of penetration of the roots of grapevine on rigosol for vine plantation on cambisol eutric on loess (Vukovo—Ilok), we found that it forms lateral branches of root in the inter-row spaces. The roots penetrate just to the depth of plowing of the cambisol (Photos 6.34, 6.35).

Guidelines for management. The guidelines are the same as for the original soil type, and so there are no

Table 6.23 Classification of rigosol

Name of taxon and base of classification	
Variety (based on use—kind of plantation)	Form (based on depth of tillage operation)
For all subtypes • Vine plantation (vitisol) • Fruit plantation • Hop plantation • Nursery • Arable land for special crops	For all varieties • Deep tilled (50–65 cm) • Medium deep tilled (65–80 cm) • Deep tilled (80–100 cm)

Photo 6.34 From the *left*: rigosols on eutric cambisol on loess (Photo: B. Komesarović)—near Osijek (P-1), on rendzina on soft limestone in Merag—island of Cres (J-1), and on kalkocambisol—Blato, Korcula Island (J-3)

Photo 6.35 *Left* The development and distribution of the roots of grapevines on eutric cambisol—Ilok (Vukovo). *Right* A model olive plantation on terra rossa—Blato (Island of Korcula)

universal guidelines and rules of thumb. The only generally important and valid guideline, as for all soils, is to avoid compaction of the soil through passing into a wet state.

6.1.6.2 Hortisol

The name of the soil type. The name comes from the Greek language (*chortos*—garden).

Table 6.24 Classification of hortisol

Name of taxon and base of classification	
Variety (based on depth of tillage)	Form (based on texture)
For all subtypes • Medium deep tilled (30–40 cm) • Deep tilled (40–50 cm)	For all varieties • Sandy • Loamy • Clayey

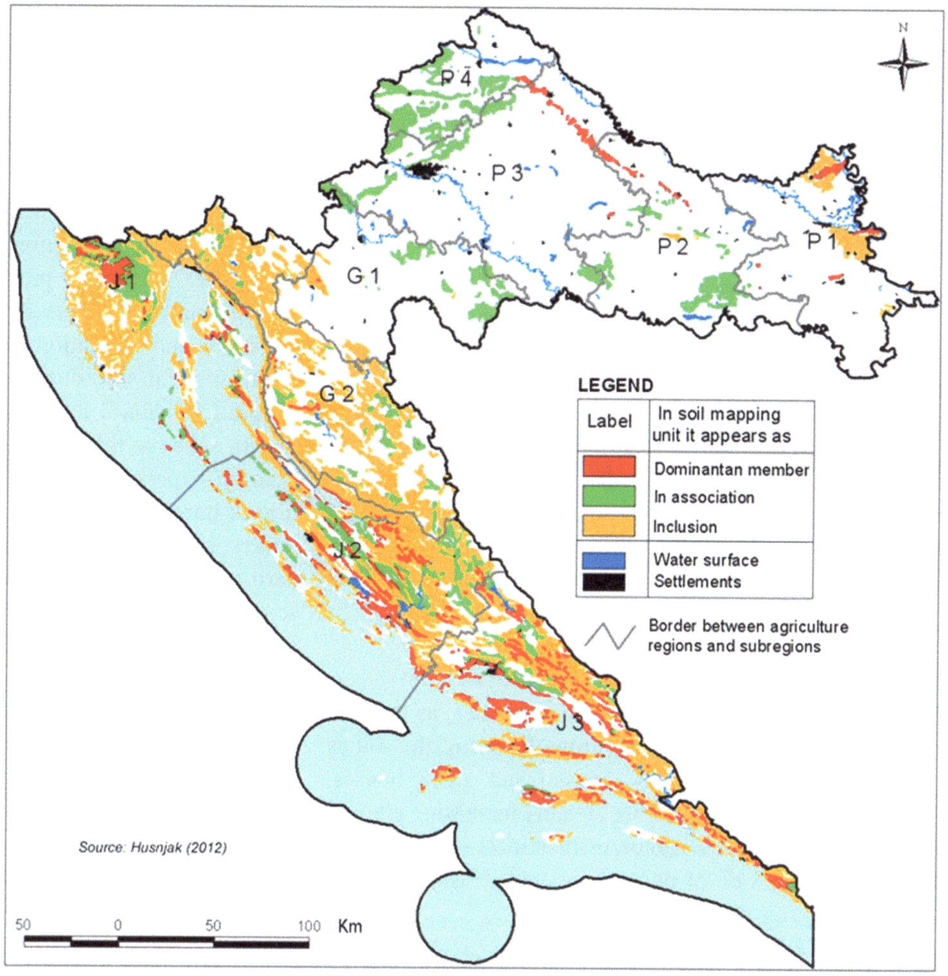

Fig. 6.19 Participation of rigosol and hortisol in the pedosphere of Croatia

Genesis and sequence of horizons. This is the same as for the already-described rigosol. This soil undergoes long-term anthropogenic influence for the growing of vegetables or flowers. Fertilization with high doses of farmyard manure or compost is common in such production.

The WRB name. Anthrosol, hortic

The unit below is divided into subtypes based on plowing depth, texture, and substance of the skeleton in the surface—P horizon (Table 6.24).

As shown in Fig. 6.19, this soil is widespread throughout the country. Unlike rigosol under permanent crops, on this soil vegetable crops are predominantly produced, which rotate much faster, almost continuously, which requires regular tillage treatment and humization by farmyard manure or other organic fertilizers, in accordance with the requirements of intensive management. The important difference in comparison with rigosol is that hortisol occurs on lower topographic positions such as in river valleys, karst fields, generally on fluvisol, and normally on soils with ground water, which is not recommended for fruit plantations. The exception to this rule is the Neretva river valley and some karst fields (Vrgoračko polje) with very intensive plantations of vine, mandarins, peach, and strawberry in the Adriatic agricultural region. The reason is the relatively high oxygen content in the underground water, the effect of which on plants is not as negative as in the Pannonian region.

Vegetables are sensitive and high-demand plants. Soils used for vegetable growing require deep, loose soil with high porosity, water-holding capacity, and aeration, all connected with the structure of the soil mainly through the high content of humus and Ca^{++} ions as a precondition for stable and favorable soil structure to the depth of the rhizosphere of vegetables.

Guidelines for management. Proper management of hortisol forms a very deep humic horizon with a higher content of humus than corresponds to the natural bio-climatic conditions in areas where there is hortisol. Maintaining a higher level of humus and care for the content of calcium are the main management aims for this soil. Abundant fertilization is one of the regular procedures (Photo 6.36).

Photo 6.36 Hortisol from fluvisol of Vrana Lake basin is a very fertile soil suitable for the cultivation of vegetable crops, particularly root, tuber, and bulb vegetables. The origin of this soil was influenced by colluvial processes

6.1.6.3 Soil of Green and Plastic Houses

The name of the soil type. The name originates, as in other soil types of the Anthrosol class, from the use of the soil as a substrate in closed spaces—plastic and green houses. Areas under plastic and glass are steadily increasing all over Croatia. The importance and permanent increase in the area of these crops is one of the reasons why Husnjak (2012) proposed this soil as a separate soil type.

Genesis and sequence of horizons. Soil under glass or plastic has a P–C/R sequence of horizons. Some of these areas are only used seasonally, such as for growing seedlings, but they are increasingly being used throughout the year for growing vegetables or flowers.

We have not studied this soil from the standpoint of the recent processes occurring in them. An exception is research on the application of cultures grown in them with respect to chemical properties, especially pH, humus content, and of course in regard to maintenance of its moisture and irrigation regime. A special problem is salinization of soil, due to the fact that these soils are not exposed to natural leaching, and are irrigated using water rich in salts, so that after some period of time salinization is to be expected, and one question is the practical question of how to proceed when this occurs. Another question relates to "tiredness of the soil," as a consequence of narrow crop rotation in a closed space. A particularly sensitive question is the optimal crop sequence and tolerance (allelopathy) of species in the breeding program (Photo 6.37).

Classification at the level of varieties and forms is made according to the criteria set out in Table 6.25.

6.1.6.4 Soil of Terraces

The name of the soil type. The name comes from the pedotechnical practice of creating flatter sections on steep slopes of erodible soils by forming "stair-like" soil surface

Photo 6.37 Soil of plastic houses on kalkocambisol—Pelješac peninsula

Table 6.25 Taxonomy of soils of green and plastic houses

Name of taxon and base of classification	
Variety (based on depth of P horizon)	Form (based on texture)
For all subtypes • Medium deep tilled (30–40 cm) • Deep tilled (40–50 cm)	For all varieties • Sandy • Loamy • Clayey

forms, called terraces. On each "step", soil is formed that has multilayer morphology, in practice a mix of the soil types that covered the slope before terracing.

Genesis and sequence of horizons. The genesis and sequence of horizons of terrace soil is generated by the building of dry walls and filling of the space within with soil material—either autochthonous or allochthonous material. The soil profile shows the following sequence of horizons: P–II–III, etc.

Our ancestors recognized a long time ago that over the entire Adriatic area, including the islands, terraces as a physical barrier provide the only efficient protection of karst soil against eolian and water erosion. Working by hand and using only donkeys for transport of "building materials", our ancestors erected a marvelous system of terraces. Every spring, using donkeys, they brought new soil which had been eroded and translocated to the bottom of slopes up to the terraces to fill the "empty space." It was a very hard job!

For some parent substrates of very favorable soils, particularly flysch and marl, the only effective and certainly the preferred method of soil conservation is provided by terracing. In spite of recent requirements and availability of machinery for terracing, today in Croatia there is much more abandonment than new building of terraces.

Almost all terraces on the islands are abandoned because of depopulation. Plant growing has moved to the karst fields or to the bottom of karst holes. An illustrative example in this regard is the islands of Korcula and Hvar. By the end of the nineteenth century the plague of phylloxera had greatly affected Korcula, which exported wine throughout the Mediterranean basin. The karst fields were cultivated for food, the bread grains—wheat and barley—in addition to the "American cultures"—potato and corn. This abandonment of terraced vineyards and the move to karst fields opened a series of sensitive environmental issues, predominantly water pollution by residues of pesticides used in viticulture. The concerns relate to drinking water and seawater (Photo 6.38).

The data on the length of dry-stone walls illustrates that the stone walls in the municipality of Blato on Korcula island alone could encircle the Earth around the equator! And this is not the only one, albeit a larger municipality of the Adriatic!

The properties of this soil type depend on the soil type on which the terraces were erected. It can be terra rossa, kalkocambisol, kalkomelanosol and rendzina on limestone, but also regosol, rigosol (vitisol) on flysch and marl and Tertiary clay and loam.

The WRB name. Anthrosol

Position in an evolutionary series. Terrace soil is multilayered and depends on which soil types participated at the time of terracing. This soil is today in practice a parent substrate for soil genesis. For example, on old, abandoned, terraces of the Šibenik hinterland we found shallow, skeletal rendzina, but on old also abandoned terraces on flysch near Grožnjan in Istria we found a developed rendzina. In every case research activities are necessary, directed at open

Photo 6.38 Terraces on Korcula have been mostly abandoned and are now silent witnesses to the persistence of our karst man. On the island of Korcula vineyards moved with them "down the field"—in the background abandoned terraces are visible

Table 6.26 Classification of terrace soil

Name of taxon and base of classification	
Variety (based on depth of P horizon (cm))	Form (based on width of terrace in m)
For all subtypes • Shallow (20–35) • Middle deep (35–50) • Deep (>50)	For all varieties • Narrow (<1.0 m) • Medium broad (1.0–2.0 m) • Broad (2.0–3.0 m) • Very broad (>3.0 m)

questions relating to the properties of these soils, to establish practical recommendations for the management of these soils in the first place. Secondly, but of no less importance, is studying the possibility of "re-activation" of old terraces. Soil on terraces is generally multilayered soil formed from soil material of different origin and properties. It is in practice a mix of soil material similar to colluvium. Research into soils on these terraces and comparison of its properties with natural soils and soils on which use has not been interrupted may no doubt provide some interesting answers to the direction of evolution of these soils on different parent substrates.

At the level of varieties and forms, terrace soil is divided according to the following criteria (Table 6.26):

Distribution. The area of this soil is not yet known.

Land use. Most of these soils as neglected or abandoned areas are used as pastures, some are afforested by natural varieties which are usually *Pinus* species. This land is ordinarily the victim of summer fires. An important land area under this soil is in full use as agricultural land for plantation of vines, olive trees, some fruits or typically Mediterranean polycultures, which in Central Istria means very broad rows of vines and some hazelnuts. On soils on flysch, traditional *Salix sp.* are grown that are used as an environmentally friendly material for vine-fastening, but in the inter-row spaces arable crops like wheat, or vegetables are grown. A real biodiversity! A very interesting topic for research is the allelopathy between different varieties, as well as varieties in agroforestry.

Walking with a home-born famous Croatian scientist in the field of vine, Prof. Mirosevic, between endless dry walls and terraces around Blato (Korcula island) on a vineyard on a terrace with evidently one owner, we registered three small houses for resting. Why three? …was my question. To save energy! …was the answer. While digging on the first third of the vineyard workers used the first one, for the middle of the vineyard the second, and for the last third the third bunya (small stone house). The main sources of energy were olive oil, wine, and fish. It was the only way to survive!

Photo 6.39 Terracing on limestone dry walls has the same role. One can see a small house within which workers took their rest, built from stone only—a bunya or kažun

Guidelines for management. These are not uniform. First, it is necessary to maintain the terrace walls to preserve their function. It is necessary to protect abandoned terraces, including afforestation in some cases, in the interest of environmental protection and as a cultural monument. On the oldest terraces it would be recommended to put up information boards with details regarding age and use and prepare a walking tour through the terraces as an example of "karst man's" fight for survival. In spite of very high requirements, the practice of terracing of soils on marl in the Pannonian region was never in use; contour tillage alone was practiced. This was a consequence of experience.

In any case, terracing is an expensive pedotechnical procedure, so only valuable crops were cultivated on them, for which the soil was normally kept in a state of high fertility. We believe that terracing will gain more importance in the future than it has today (Photo 6.40).

Photo 6.40 On the erodible soils under Kalnik hill many years of contour plowing gradually form grassed terraces

6.1.6.5 Karst Anthrosol

The name of the soil type. The name of this soil is composed of two words: karst as a special geolithological formation and anthrosol, which means soil as the creation of man.

Genesis. Raising permanent crops on substrates obtained by grinding limestone in the Adriatic agricultural regions is becoming a more widespread practice. There are several methods of preparing such ground:

- manual or mechanized cleaning/removal of stone/rock;
- mechanized removal of stone deposited on the soil surface;
- mechanized removal of stone, fragmentation by rotavator on the surface;
- grinding of stone in situ, mixing with soil and natural vegetation and depositing on the surface.

Table 6.27 Taxonomy of karst anthrosol

Name of taxon and base of classification	
Variety (based on original soil type)	Form (based on stone content)
For all subtypes • Lithosol • Kalkomelanosol • Rendzina • Terra rossa • Kalkomelanosol	For all varieties • Low content < 15 % • Medium stone content 15–30 % • Rather more stone content 30–50 %

There are of course various combinations of these procedures. The technical features of this process are significant, because it requires machines that can easily cut and grind the roughest stone, vegetation and soil, and arrange it

Photo 6.41 The first karst anthrosol in Croatia from rendzina on soft limestone—vineyard of Badel, Zagreb, on the site Korlat (J-2) at Benkovac-Zadar. The material contains much soil mass, but it is unevenly distributed as can be seen on the *right*, and its redistribution has been done recently. The result is uneven distribution, which requires selective treatment of certain sites/parts of the vineyard, which is not simple

Photo 6.42 Humans make their own decisions (*left*), but Nature does not accept wrong decisions and makes its own in return. Erosion (*right*) shows the importance of the correct direction of rows. It would be better to use contour direction of rows

as a substrate whose features depend mostly on the material from which it was prepared. If it is a soft limestone, which in addition contains a lot of soil rich in humus, it can be prepared as a suitable substrate for the establishment of plantations. Such a case is shown in Photo 6.41. However, as seen in Photo 6.39, the natural topography of the terrain caused by the differences remains even in old plantations.

The WRB name. Anthrosol

Position in an evolutionary series. Karst anthrosol is held in a state favorable for the cultivation of selected crops or vegetables. The evolution of the soil to a considerable extent depends on the content of the soil skeleton and the depth. If erosion is prevalents the skeleton will certainly affect the processes in the soil, or the redistribution of soil material. When and if there is a complete natural succession between the rows of root crops that will slow the erosion of time and if there are stable relief positions, it is to be expected that the processes of pedogenesis will start by creating the initial soil (A)–C–R sequence of horizons in the profile, which will then slowly develop into a humus-accumulative A–C–R soil type, i.e., rendzina. The division of the lower taxons is made according to the following criteria (Table 6.27):

Distribution. These soils are completely "new members" of the Croatian pedosphere, so they occupy a very small area, not more than 1000 ha.

Guidelines for management. All the features and guidelines for the management of this soil depend on the original soil type, content and the relation between skeleton and soil mass features of the slope, etc. Erosion control is certainly the most important, and apart from rows following the direction of contour lines it requires the planting of grass between the rows.

The preparation of karst anthrosol and of equal importance the management of the plantation have no generally applicable rules or universal patterns. Preparation should be under the full control and supervision of professionals. It is our opinion that in addition to the problems already raised in regard to plantations time will bring with it more problems and raise new questions. Problems will also increase due to the relationships of soil material with the "bottom" of the substrate, and its redistribution depends on "the underground relief" which is very diverse.

The direction of rows of plantation should be parallel with the contour lines. These procedures should not be considered acceptable in positions with little soil, no soil, or on slopes or notches at the base of coastal and island heights, for example, below Biokovo. What are the problems? "The technical side" has a solution: machines can do everything; however, Nature may not be ready to follow the will of man, so problems arise in the exploitation phase (Photo 6.42).

6.1.7 Class VII: Technogenic Soils (Technosols) I–II–III–IV…etc. Layers

6.1.7.1 Deposol

The name of the soil type. The name is based on the origin of this soil, which means deposition of waste material or soil mixed with parent material of different properties and depth.

Genesis and sequence of horizons. Deposition of different materials from urban areas on flat relief or on variously formed depressions results in the formation of this soil. It should be noted that at the beginning this is chaotically distributed material which is not a soil, but a very short time after its stabilization we see the first appearance of weed vegetation adapted to such habitats. The residue of this vegetation mass forms the initial (A) horizon, followed by a humus-accumulative A horizon. Chemical decomposition from deposited waste is a normal process resulting in the forming of humus from organic waste, so unchanged waste

Photo 6.43 A common method of formation of barren soil near Zagreb. We can see a site where there has been "wild" removal of gravel followed by "wild" disposal of waste. These different waste materials will start to form a soil, thus covering up the illegal practices

Photo 6.45 Phosphogypsum stack of Fertilizer Plant—Kutina

has the function of a C horizon. The vegetation mass is very high and will increase further if not covered by new waste. Of course, it is possible for soil material from building activities to be deposited. In this case the processes are the same and an initial (A) horizon is followed by the A horizon, and leaching and cambic processes.

Position in an evolutionary series. The vegetation is very dense and relatively quickly forms the first initial (A), then a humus-accumulative A horizon.

The WRB name. Anthrosol

Distribution. The distribution of this soil is officially very small, but there are numerous "wild" landfills. The largest localities are areas of disposal from specific waste generators like the thermal power plant Plomin with coal ash as waste material, which is gradually stabilizing, and the petrochemical plant in Kutina city with its phosphogypsum disposal site in which the material is still dispersed, and in one area will soon begin stabilization and greening (Photo 6.43).

Physical and chemical properties. These are not homogeneous because of the heterogeneity of the disposed waste material (Photo 6.44).

Phosphogypsum is a special waste material of the fertilizer industry in Kutina that is hydraulic disposed in a water-impermeable stack. Water is constantly circulated to neutralize its low pH and releases fluoride using lime. Special tests were required to identify plant species and substrates, for recultivation and greening of stack, that are capable of surviving these conditions. In 4 years of field experiments we find a mix of *Amorpha fruticosa*, alfalfa *(M. sativa)*, *Trifolium alba* and *lupulina*, *Cynodon dactylon*, and *Festuca rubra* on a clay layer (5 cm) as the best

Photo 6.44 *Left* Plant coal ash pulp—Plaski (G-1). *Right* Coal ash from the coke plant in Bakar (J-1). Vegetation already covers the surface of the landfills

Photo 6.46 Among the several plant species involved in the experiment, alfalfa (*Medicago sativa*) (*left*) and Amorpha (*Amorpha fruticosa*) proved to be suitable species for greening the landfill (*right*)

solution for green cover (Gašpar et al. 2009, Leaković et al. 2012) (Photos 6.45, 6.46).

The question of how to use deposol is a delicate issue, especially because the deposited material may contain harmful substances such as heavy metals and toxic elements, which can enter the food chain if these plants are used by domestic animals, wildlife, or even through honey via honey plants which can grow on such soils.

The classification of Croatian soils also includes soil flotation sludge and atmospheric deposition on soil. There is very limited information on these soils.

6.2 Order of Semiterrestrial Soils

6.2.1 Class I: Soils with Stagnation of Rainwater of A–E_g–S–C Soil Horizon Sequence

The name of this class is derived from the Latin *stagnare*—stagnate, and describes the main characteristic of the soils, the stagnating of the surface—mainly through rainwater precipitation, which stagnates on a compacted soil horizon of low permeability. Such wetting, a type of hydromorphism, occurs in a climate with relatively high annual rainfall. A common feature in all the soils of this order is that the water (its surplus and/or deficit) is the main factor that limits production in agriculture and determines the natural plant species adapted to these conditions. For use in agriculture, the water should be removed from the soil by hydromelioration—drainage operations.

6.2.1.1 Pseudogley

The name of the soil type. Named by the famous German soil scientist Kubiena and composed of the words; *pseudo*—false and gley (from Russian glei—mud)—marshy soil, to distinguish it from "real" existing gley with hydromorphism caused by underground and/or floodwater, or "water aside." In the older literature, it is described as podzol, later called parapodzol, or surface gleyed soil. People call this soil "silty soil" because of its high content of silt.

Genesis and sequence of horizons. There are two forms of pseudogley genesis. One is the process of clay particle migration via eluviation–illuviation to the stage at which a horizon formed by illuviation enriched with clay becomes impermeable to rainwater. The second path is the appearance of rainwater-impermeable Pleistocene loams altered by translocation via glacial processes, which were at the end of the Pleistocene and/or early Holocene covered by "fresh" and more permeable material of eolian origin, described as "loess-like" material. The typical endomorphology of the sequence of horizons is A–E_g–S–C. The substrates upon and in which the characteristic morphology of pseudogley forms may be of very different origin: eolian, fluvial, and colluvial deposits but in the Middle Slavonian hills and Moslavačka gora through the physical weathering of rock into smaller fragments—a regolith of siltstone. In all cases, in the surface layer the impermeable horizon leads to an alternation of two extreme conditions. After rain there is the wet phase, in which soil is completely saturated with rainwater; there is no oxygen in the soil, so anaerobic conditions and processes of reduction are prevalent. After evapotranspiration comes a dry phase, in which processes of

Photo 6.47 Interior of marmoral peds—part of the soil matrix of the S horizon of pseudogley, with a mixture of reddish (oxidized) and brighter (reduced) zones and dark, soft concretions

oxidation dominate. As the soil material is compacted and impermeable, compounds formed by reduction remain in place and diffuse laterally, creating a bluish-gray zone, within which the compounds are reduced, and they form a rusty-red-colored zone, within which there are oxidized compounds. The acid chemistry conditions and the strong reduction of iron and manganese allow them to become more soluble and mobile and they can bind to soluble low-molecular-weight acids from the humus forming dark "ball-like" forms called humus–manganese concretions. The change from rusty to grayish-blue zones in the S horizon of the profile leads to the typical marmoreal appearance. These processes result in specific redoximorphic parts of the soil matrix, as visible in Photo 6.47.

The genesis of pseudogley is complex and related to numerous specific climatic, geological, relief and vegetation related processes, based on the conditions at the time. In recent times, these conditions have been semihumid and humid climate, with over 700 mm of rainfall annually. For water stagnation, relief conditions are required that hinder or impede rapid lateral runoff. These are slightly undulating, hilly areas of Pleistocene terrace, up to 600 m above sea level. The parent substrate has a very important role in the forming of pseudogley, and most are loamy sediments of Pleistocene non-calcareous material (loess leached and changed by fluvioglacial moving) as described in a separate section, alluvial deposits, and Tertiary clayey sediments, etc. The basic requirement for formation of a pseudogley under a relatively permeable top layer is development of a rainwater-impermeable layer on which water stagnates and gradually fills the layer above to the surface. This layer may be of geological origin—in this case pseudogleyization occurs as a primary process. It rarely occurs over marls, but especially over clay. Natural vegetation in pseudogley areas

Table 6.28 Taxonomy of deposol

Name of taxon and base of classification	
Variety (based on original soil type)	Form (based on depth of material)
Younger soils—without (A) horizon	Shallow 30–50 cm
	Medium deep 50–100 cm
Older soils	Deep 100–150 cm
	Very deep > 150 cm

Table 6.29 Taxonomy of pseudogley at the level of variety and form

Name of taxon and base of taxonomy	
Variety (based on depth of S horizon)	Form (based on pH)
For all subtypes	For all varieties
Shallow (<25 cm)	Form (based on pH)
Medium deep (25–50 cm)	Eutric (pH > 5. 5)
Deep (>50 cm)	Dystric (pH > 5.5)

are oak and hornbeam (*Querco-Carpinetum croaticum*) forests (Vukelić and Rauš 1998; Vukelić et al. 2008). Tannins and other resins secreted by roots of oak increase the solubility and migration of iron in soil. Classification of lower units of the type is performed based on the relief position in which it occurs, impermeable horizon depth, and degree of saturation (Table 6.28).

Position in an evolutionary series. Pseudogley is the last member of a pedo-evolutionary series on Pleistocene sediments, starting with regosol, and continuing with cambisol eutric and luvisol on loess. The zone of pseudogley continues naturally after the zone of luvisol and under certain conditions occurs in association with these soils, forming as the next evolutionary stage of luvisol pseudogleyic.

If our observations and thesis on irreversible coagulation of some mineral colloids in the S horizon of pseudogley in an excessive hot (and dry) summer, similar to the formation of the stable soil structure of terra rossa and kalkocambisol, is confirmed, then this is also a logical hypothesis for a possible retrograde–reversible direction of pedogenesis of pseudogley. Namely, after a hypothetical coagulation pseudogley would be in terms of permeability similar to luvisol. This means that after a new extreme warming event the S and/or B_t horizons will again become water permeable! The taxonomy of pseudogley is based on criteria that influence soil fertility: depth and chemistry of the soil (Table 6.29).

The deeper varieties are favorable because the duration of the ecologically unfavorable dry and wet stages is shorter than in the shallow variety. And, of course, the eutric form is favorable because it is richer in bases and nutrients, and has a higher CEC and other chemical properties, including humus quality.

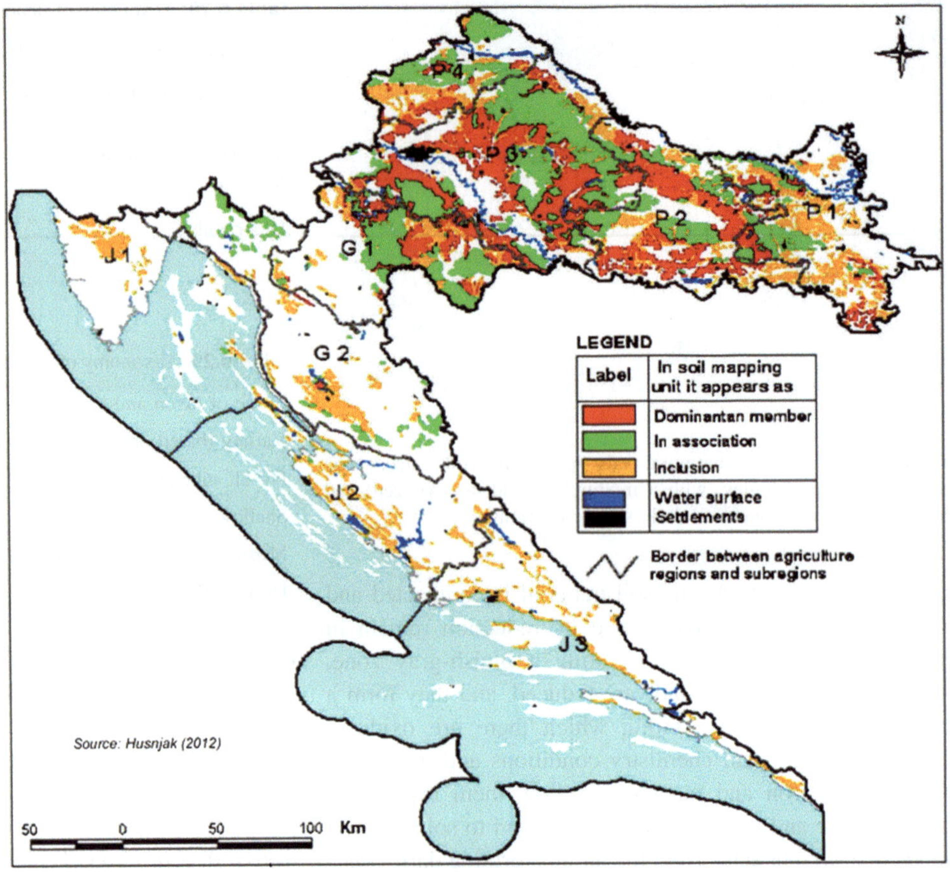

Fig. 6.20 Distribution of pseudogley in the pedosphere of Croatia

The WRB name. Stagnosol, partly albeluvisol

Distribution. With 577,025 ha pseudogley occupies 10.4 % of land territory as the second most widely spread soil type according to the distribution in the pedosphere of Croatia (Fig. 6.20).

This map unit occurs predominantly in the Pannonian and Mountain regions, including all subregions, as inclusions, and in the Adriatic region, at positions related to Pleistocene eolian deposition. The process of pseudogleyization is possible, and terra rossa, particularly the relict variety, if compacted, has become water-impermeable. At its most easterly location it occurs as inclusions or in association with gley soils in the Spacva forest. To the west, it mostly occurs on the Pleistocene plateau in the valley of the Sava and Drava River basin, including the southern slopes of the mountains and foothills of the Middle Slavonian hills, and Moslavacka gora and Medvednica at the border with Slovenia (Photo 6.48). On the southern slopes of Bilogora in Donje Međimurje it is found on the Pleistocene plateau of the Drava River in association with other, mostly luvisol, soils.

Plateau pseudogley occurs on Pleistocene deposits where relief conditions prevent lateral flow of rainwater, so that there is a long period of stagnation in the S horizons an impermeable layer. Depending on the rainfall quantity and the depth of the impermeable horizon, in small depressions rainwater can escape to the surface or be visible and found in channels, as shown in Photo 6.49.

Pseudogley of slope is expressed by lateral flows and runoff of rainwater, slowly flowing on the impermeable horizon under the influence of gravity in the direction of the slope. In places where the impermeable horizon emerges at the surface the water can escape in the form of short-term sources—so-called "pištavac." According to the depth of the impermeable S horizon, which is ecologically very important, all subtypes are divided using the criteria of Table 6.28. Depending on the chemistry and the degree of base saturation of the CE complex, we distinguish dystric and eutric pseudogley.

Physical and chemical properties. Pseudogley is a soil type of extremely unfavorable physical properties; several times a year, sometimes also during the vegetation period, the impermeable S horizon causes adverse changes with its phytoecologically unfavorable wet phase, where the soil is saturated with excess water, and then we have dry phases where deficit of water dominates (Table 6.30). In the saturated phase it is nonstructured, clayey and shows a preference for erosion, but conversely, in the dry phase, it is compact and also nonstructured. There is a very short interval of wetness favorable for tillage. Over a long period in the spring pseudogley is cold and unfavorable for any

Photo 6.48 Pseudogley of a Pleistocene plateau—Mihaljevec in Gornje Medjimurje (P-4 subregion)

tillage. Finally, pseudogley has a shallow physiologically active depth for root penetration, because the S horizon is too compact to allow penetration of roots.

As can be seen, there is a marked textural differentiation between the S and top P horizons. In addition, the high silt fraction content in the soil causes an unstable structure, crust formation tendency, high runoff and erodibility. If the soil is left bare even only for a short time and raindrops hit it, this may cause buildup of a crust and surface runoff, which paves the way for erosion (Photo 6.50).

The chemical properties of pseudogley are also unfavorable. It is acid to strongly acid soil, non-calcareous over its whole depth. In very acid pseudogley, Al^{+++} ions may be freed in solution or adsorbed in CE complex at phytotoxic quantities. It has a low CEC and base saturation level of 30–60 %, dominated by H^{+}, Al^{+++}, and Fe^{+++} ions. The affordable supply of nutrients is highly variable, nitrogen and phosphorus is very low and potassium in low to moderately low. Because of the unfavorable physical and chemical properties, the biological activity of pseudogley is low; the prevalent microorganisms are fungi and anaerobic bacteria. In the wet phase possible adverse negative denitrification processes may occur.

Land use. Land with pseudogley is used as a quality habitat for valuable forests of sessile oak and hornbeam. On favorable relief positions, pseudogley is used in agricultural production. The yields of nonameliorated land are low, unstable, and very dependent on climatic conditions, i.e., in dry years the crop is damaged by drought but in wet years it suffers from excess water from the abundancy of rainwater.

Guidelines for amelioration and management. In relation to environmental characteristics pseudogley is a very

Photo 6.49 Pseudogley on a plateau: rainwater escaping out onto the surface, is firstly visible in canals (*left*), where, if it lasts more than a few days, it can damage crops as shown in the *right* photo

Table 6.30 Physical and chemical properties of pseudogley

Soil horizon	pH (nKCl)			mg P_2O_5/100 g of soil			mg K_2O/100 g of soil			Humus (%)			% of clay		
	Average	Minimum	Maximum	Average	Minimum	Maximum	Average	Minimum	Maximum	Average	Minimum	Maximum	Average	Minimum	Maximum
Pseudogley of plateau (terrace)															
P	4.9	3.7	5.9	3.5	1.8	9.2	8.4	4.0	18.2	2.3	1.3	4.4	17.6	10.4	27.6
E_g	4.6	3.7	6.0	1.7	0.3	4.0	6.6	3.2	9.2	0.7	0.3	1.1	24.9	14.8	36.1
S	4.6	3.7	6.4	–	–	–	–	–	–	–	–	–	26.6	20.9	40.3
C_g	5.0	3.7	7.3	–	–	–	–	–	–	–	–	–	23.9	13.6	35.2
Pseudogley of slope															
P	4.8	3.8	6.0	1.9	0.6	4.2	11.1	5.0	20.4	3.2	1.8	8.3	17.1	9.4	23.4
E_g	4.5	3.9	5.6	0.6	0.0	2.0	6.7	3.1	10.2	0.9	0.2	1.9	21.7	14.6	30.4
S	4.3	3.6	5.4	0.4	0.1	0.8	5.3	3.7	8.4	–	–	–	23.5	17.4	32.2
C_g	4.6	3.6	6.0	–	–	–	–	–	–	–	–	–	23.7	16.4	31.0

Source Bašić et al. (2003)

Photo 6.50 The high content of silt in pseudogley causes crust formation and creates a tendency for soil erosion, especially if the soil is tilled and left bare with no crops—Blagorodovac in Daruvar (photo: Kisić)

heterogeneous unit but the natural fertility of pseudogley is mostly very low due to its unfavorable physical and chemical properties. Therefore, for its use in agriculture, for high, stable yields and to grow high-demand crops, it is obvious that amelioration is required.

From an ecological point of view, two very important properties are depth of the impermeable layer and position of the pseudogley in the relief. Slope pseudogley is ecologically favorable with its deep impermeable horizon. What is the effect of the deeper S horizon? The pseudogley may receive and retain a larger quantity of rainwater without saturation, risk to crops, and anaerobiosis. The water regime is different in lower, middle, and upper parts of slopes. On the upper part of the slopes rainwater drains and moves relatively quickly, and the damaging wet phase is short; in the middle of the slope it is a little longer, and the longest duration of the wet phase is seen in the lower part of slopes—at the bottom. Extreme water regime conditions are generally more pronounced in the shallow variety of pseudogley, where even a small amount of rainfall causes the wet phase, but drying out through evaporation and transpiration is also very fast. The chemical properties of pseudogley are very heterogeneous. Eutric pseudogley has a higher content of bases and is more favorable.

To correct the unfavorable natural production characteristics of pseudogley it is necessary to implement amelioration procedures. The risk of excess storm water can be reduced by deep plowing, i.e., deepening the plow layer at the expense of the impermeable horizon. This increases the

Photo 6.51 Moving water flowing on the impermeable S horizon after subsoiling (*left*). If water is not removed from the plot, machinery will sink into the subsoiled pseudogley to a depth equal to the impermeable horizon (*right*)

water-holding capacity in the plow layer, i.e., the possibility of holding larger quantities of rainwater. This procedure satisfies conditions of relatively low rainfall in the case of a natural slope to drain excess water. It is believed in Croatia that deep plowing can solve the problem of excessive rainwater that does not exceed 50 mm. Higher quantities of rainfall require the removal of sufficient water by drainage (Photo 6.51). A common approach for a melioration of pseudogley is called threefold or integral land reclamation. This consists of three interrelated phases: the first is evacuation of sufficient water by installing pipe drainage in the soil, the second is subsoiling or loosening the soil above the depth of the drain pipes, and the third is liming with humification and ameliorative fertilization. Drainage is tasked to evacuate excess water into the collector network—open canals. Because of the high content of silt on the drainage pipes they are usually covered by a layer of filtration material (gravel or plastic pellets) called a hydraulic filter, which serves as a contact layer between the soil and drain pipes. Sufficient water moves through the loosened mass of soil to the gravel and drops into perforated pipe drains, which take it to the collector network. An alternative is open canals, where open drains enter the hydraulic filter material. The purpose of subsoiling, loosening, or mole drainage is to allow the inflow of water into drainage pipes. These procedures are called secondary treatment (primary treatment is the installation of pipe drainage). Liming and humification is a form of amelioration aimed at creating a stable structure that increases infiltration of rainwater, prevents surface runoff and erosion, and allows the inflow of water through loosened, subsoiled soil or through mole drains and drainage pipes. Agrotechnical amelioration is necessary to address other unfavorable properties of pseudogley. Liming and fertilization using high rates of mineral fertilizers, and plowing and humification of plant residues are both required with deep plowing. A deep layer of physiologically active topsoil can achieve these operations: improving the structure, air–water relationship, and chemical and biological properties of soil. Deep plowing and liming should be regularly repeated every few years. This complex of ameliorative treatments to resolve all constraints on plant growth may achieve very high and stable yields of most annual agricultural crops, and it is possible to grow a wide variety of crops. However, access to pseudogley a melioration, because of the heterogeneity of this pedosystematic unit, may not be uniform. In each case it is necessary to carry out appropriate preliminary studies whose results should guide the approach to a melioration. Nonameliorated pseudogley in Croatia has a very valuable potential for increasing the agricultural area. After a melioration, the pseudogley is favorable for the cultivation of not only arable crops but also vegetable crops and fruit production. This soil requires energetic soil conservation interventions, especially conservation tillage, because of the high soil erodibility of pseudogley. From the standpoint of protecting pseudogley soil from erosion, a high participation of row crops, especially maize, creates a disadvantage. The cultivation of this crop usually overlaps the period of greatest risk of erosion. This is the stage of seed-bed preparation and sowing of spring crops to germination (Photo 6.52). Until the stage of crop development is reached where inter-row space is closed by crop mass, spring rains of high intensity cause high erosion. Large amounts of residues of herbicides and other agrochemicals are deposited in the accumulations of erosional sediment at the foot of the slopes. Eutrophication and/or pollution of aquatic ecosystems may occur if these enter the watercourses.

Photo 6.52 Just after seed-bed preparation and seeding of maize there was abundant rainfall, which moves soil enriched with nutrients and herbicides down the slope and accumulates it at the foot of the slopes—Puljkovac Garesnica (P-3)

Therefore, the cultivation of arable crops should be practiced with conservation tillage, and one should strive to avoid or shorten the maximum period of bare soil. What remains as a permanent injunction for the management of this soil is the principle that every transition into an unfavorable state of moisture, i.e., the soil being too wet, always brings more harm than good and therefore could not possibly be advisable. Key management goals are a high content of humus and microbial activity, which is why they require an adequate covering of livestock units per ha of arable land. A favorable circumstance is that these soils are so pure that one can practice and develop all forms of ecological agriculture.

Finally, let us note that even on this soil after performance of the integral amelioration treatments described over several years it is possible to carry out reduced tillage and no-tillage according to the concept proposed by Butorac (1999): an initial maximum (investment in amelioration) and than a minimum (tillage). Our experience with 15 years of no-tillage on pseudogley of a slope in Daruvar (Kisić et al. 2005, 2006; Kisić and Bašić 2012) favors this practice. In the first few years, yields were lower (but also production costs), and then increased over 15 years to be now fully equal to the yield of an experimental variant with conventional tillage.

6.2.2 Class II: Anthrosols Semiterrestrial P–S–C Soil Horizon Sequence

In this class are semiterrestrial soils, whose properties are substantially altered during acid soil amelioration projects. These procedures include deep plowing—plowing or deep loosening—complex pipe drainage procedures. Hydroamelioration of the soil by drainage radically changes all its characteristics, because the drainage stimulates oxidative processes in the soil, although the soil retains its original morphology long after the performance of land reclamation.

6.2.2.1 Rigosol

The name of the soil type. The name comes from the very deep plowing (rigol) carried out as the primary tillage operation, which forms an anthropogenic P horizon and causes a radical change in endomorphology—number, color, and sequence of horizons, and gradually all properties. Namely, plowing takes A–E_g and part of the S horizon, mixes all the materials and forms a new one—P, different from the original horizons (Photo 6.53).

Physical and chemical properties. The physical and chemical properties are significantly altered in relation to the underlying soil type and under the direct influence of man and the farming system. The plowing produces a great weight of "row" soil—virgin soil, which requires regular fertilization with organic manure and liming, to establish and maintain the best possible biological activity.

Land use and guidelines for management. This soil is ameliorated only for use in intensive crop cultivation and/or vegetable production. Subtypes of pseudogley can be used for orchards. These soils are not allowed to transition into the wet phase. This is a general and always valid rule. There is also an obligation to repeat all measures—trenching, liming, and fertilization.

Photo 6.53 *Left* Rigosol originating from pseudogley—Gradiška (P-2). Visible is the depth of trenching at over 70 cm. The P horizon is changed by humization, liming, fertilization. *Right* Without sufficient water removal by pipe drainage the roots of Trapper from the winter period in a water saturated P horizon leave traces of boots of walker from winter period

6.2.2.2 Drained Pseudogley

This type of soil occurs through drainage of pseudogley performed by so-called integral or three-stage drainage. The first step is the installation of drainage pipe—perforated drainage pipes that are tasked with accepting water from the upper layers and transporting it to the drainage network. The drainage pipes are covered with a layer of gravel or perlite, which represent the so-called hydraulic filters, protecting the drainage pipes from blockage of the pipe perforation by silty mud from the soil suspension inflowing into the pipe. The second stage is deep loosening, which runs perpendicular to the direction the drainage pipes are laid, to allow inflow of surface water into the hydraulic filters, which after filtering drops into the drainage pipes (Fig. 6.96, Photo 6.54).

The third step is soil treatment and creation of a favorable structure in the top layer, the P horizon, by liming, as Ca^{++} ions perform the function of coagulating colloids, forming microaggregates. Humization follows because humus is the best material for the cementing and forming of microaggregates, which entails a stable and favorable soil structure. None of the treatments described is expensive, and hydroameliorated pseudogley is a very valuable soil type. During the stage of full exploitation, i.e., in the postmeliorated period, the system requires professional management and maintenance of the ameliorative system (Vidaček 1998). This includes restoration of the loose fabric, regular liming, and continuous care of the humus content and "turnover" of humus in the soil (Butorac et al. 1988). It remains to state again as a permanent limitation that the soil should not be allowed to transform into wet soil, which can only be defined as harmful. Experience in relation to this point is rich and very supportive of this statement.

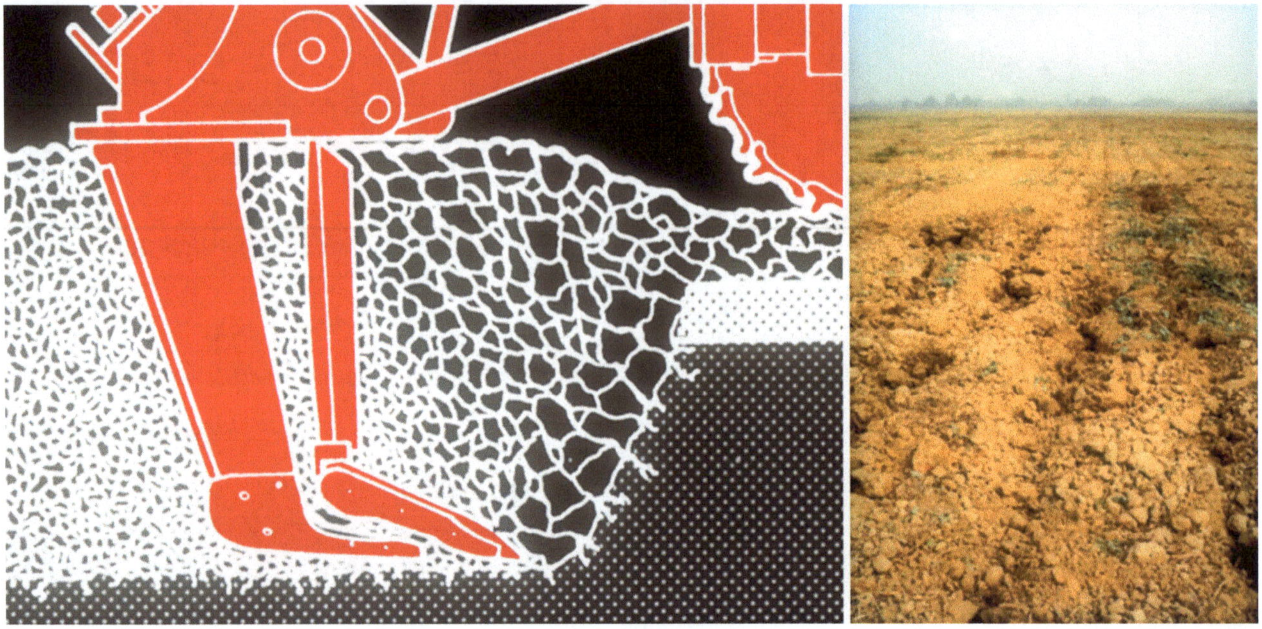

Photo 6.54 *Left* Scheme of so-called deep loosening by an "active deep loosener" which has a working body that penetrates into the S horizon, vibrates in all directions and breaks the soil mass of the pseudogley, opening the path for inflow to drainage pipes. *Right* Surface of subsoiled pseudogley

6.3 Order of Hydromorphic Soils

6.3.1 Class I: Initial Multilayer Soils of I–II–III–IV…$g_{r/ox}$

6.3.1.1 Fluvisol

The name of the soil type. The name is of Latin origin (*fluvius*—river). The name fluvisol includes recent, as a rule multilayered, river, marine, and lake sediments, in which any other process of soil genesis is usually minimal because the dominant process is sedimentation and deposition.

Fluvisol is a globally recognized, young azonal soil. The first great civilizations germinated on this soil on the banks of legendary rivers such as the Nile, Euphrates, Tigris, Indus, and Danube, including the Vucedol civilization in Croatia (Juric 2011). The mystery of the fertility of fluvisol along the Nile motivated all Egyptians to probe the ever-present question of the origin of life in "al-quimya," which in ancient Egyptian means investigation of the black soil from which germinates life. This was the origin of alchemy and, of course, later modern chemistry. The ancient name for Egypt was Kemu—which means black soil.

Genesis and sequence of horizons. This is similar to the class of undeveloped automorphic soil. The initial humus-accumulative (A) horizon of fluvisol is formed by deposition by the river. The soil is a genetically young, recent soil with distinct layers of river sediment containing clay and humus; thus, it has a natural fertility, but only after deposition of the said materials (Photo 6.55).

The deposition of material in transit in the flowing water occurs at the moment the transmission power of water matches gravity. At this point the material is selectively deposited, first the heavier material, followed by the lighter and finally the colloidal fraction—clay and humus. The deposition of materials occurs very regularly with each flood, but as flood intensity has historically changed, and the course of river beds has also changed, the timing of recent material depositions has been rather chaotic. The profile is layered, and deposition of new material over "old" prevents the formation of a horizon and the soil profile has an (A)-II–III–IV… etc. sequence.

The WRB name. Fluvisol

Position in an evolutionary series. Its position is a specific one. The dominant process is sedimentation of a new soil layer. Of course, in spite of bringing in new soil material of high fertility, the flooding of fluvisol is not a desirable process, because it limits the growth of all crops living on the soil during the flood period. After flood protection, new soil genesis begins, in which fluvisol becomes the parent substrate for soil genesis under changed, automorphic conditions. Soil genesis in this case starts with a soil of (A)–C profile, which means regosol on fluvial deposits, followed by automorphic soil of A–C class, i.e.,

Photo 6.55 A typical Fluvisol with sequence of soil horizons: P–II–III–IV on the right bank of the Drava River—Križovljangrad near Varazdin (P-4 subregion). The soil is light, sandy and loose, of very favorable, practically optimal properties

rendzina or ranker (leptosol, eutric, or dystric), then, in the far future, cambisol, etc. The criteria of taxonomy of fluvisol in soil units below subtype is shown in Table 6.31.

Distribution. Fluvisol in Croatia is related to the larger rivers. With the largest rivers, the Sava, Drava, Danube, and Neretva rivers, karst fields occur as the dominant pedosystematic unit at 136,343 ha or 2.5 % of land area (Fig. 6.21).

On the floodplains of rivers, but also other, smaller streams, it occurs in the form of an inclusion. Narrow river valleys, especially in the Mountainous region (G-1 and G-2), transfer the soil material by fluviative and colluvial processes, which vary in dominance. This results in a mixture of various materials, including stone. During the transfer of fluvial material, if abrasion occurs, the stone becomes rounded and transformed into gravel. Conversely, colluvium has a shorter "path" to sedimentation and it is characteristically sharply angular and nonselective stone, which is one of the features for identification of soil of different sources. Of course, some colluvial material can be calcareous, eutric, or dystric, and others not.

Table 6.31 Taxonomy of Fluvisol

Name of taxon and base of classification	
Variety (based on depth to the gravel)	Form (based on texture of fine soil)
For all of the subtypes • Very shallow > 15 cm • Shallow 15–30 cm • Moderately deep 30–50 cm • Deep 50–90 cm • Very deep < 90 cm	For all of varieties • Sandy • Loamy • Clayey

Let us stress that fluvisol has enormous importance, far greater than could be inferred from its area, because it represents one of our most fertile soils on which is developed very intensive agricultural production. Its importance is also seen in the fact that the origins of world civilizations, including Vučedol (near Vukovar), are tied to this soil type. Today we know that at the time the Vučedol civilization flowered, the Vukovar loess plateau was a steppe pasture full of wildlife with locally shallow chernozem (Photo 6.56).

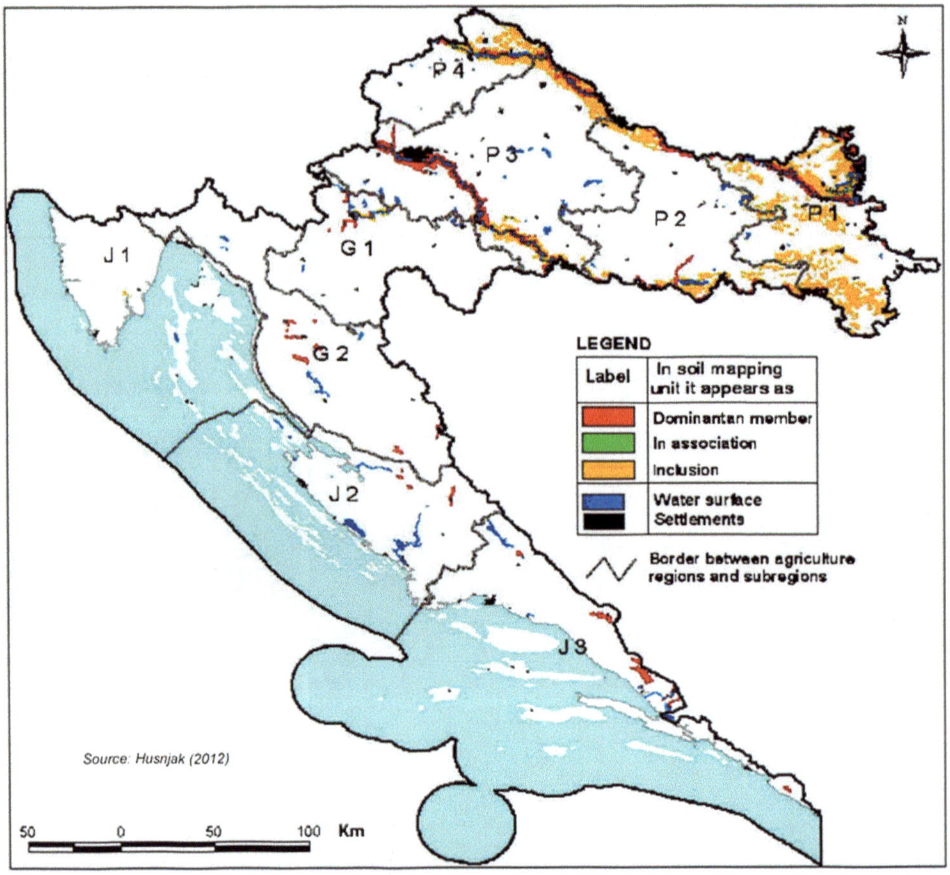

Fig. 6.21 Distribution of fluvisol in the pedosphere of Croatia

Physical and chemical properties. The physical and chemical properties of fluvisol depend on the type of material and conditions of transfer, sedimentation, and deposition, as well as on the texture and other physical features. The texture is very heterogeneous, from skeletal to clayey. Fertility of the clayey and skeletal form is the lowest, because of lower water holding capacity, CEC and plant nutrients supply in skeletal, but heavy, sold and hard soil in clayey form. As a rule, most fluvisols have favorable physical properties. Chemical properties also depend on the type of sedimented and deposited material, but also on the groundwater level and its chemical properties and alteration level. The soil can be non-calcareous or calcareous, and if groundwater is alkaline or salinized the soil can therefore have a weakly alkaline to weakly acid chemistry, and be salinized or alkalized. The content of humus in these soils is highly variable and generally lower in the skeletal and sandy compared with other forms of fluvisols. It is not uncommon for the deeper layers to contain more humus than the top layer. Natural available plant nutrient supply is generally variable, and depends on the type of sedimented soil materials.

Land use. This depends firstly on properties and the water regime—the number and periods of annual floods. Often the groundwater rises through the capillary system to the rhizosphere, and thus establishes an optimal supply of water for crops. Such fluvisols are extremely valuable agricultural land. They are used for growing high-demand arable and especially vegetable crops. The disadvantage is that the usually unpredictable floods may restrict the choice of crops; for example, it may not be possible to produce winter crops because of damage by floods. Thus, in the Sava River valley basin (at Jasenovac and to the east of Sava River valley) what are known as "everlasting corn fields" have appeared, where monoculture farming is practiced leading to very high returns. The fluvisol is of special significance, because it supplies cities located in river valleys with vegetables. High-intensity vegetable production is a good choice, because vegetable production can bring a return on the investment in amelioration—the irrigation of these soils.

Guidelines for management. Management of fluvial soils depends on the one hand on the physical, chemical, and biological characteristics of the soil, and the intended farming system and program on the other. Where amelio-

Photo 6.56 *Left* Valley of the Danube River in Opatovac, close to archeological finds related to the Vučedol civilization. *Right* River island—Ilok ada with fluvisols below Ilok city (P-1)

ration treatments are concerned, the main limiting factors in the cultivation of most fluvisols is the risk of flooding. Another is the requirement for irrigation due to the skeletal or sand content, because of the low water-holding capacity of sandy and skeletal forms and the frequent lack of water in the summer. If we are producing vegetables then the risk is too high when comparing the value of production to the costs of the irrigation system. Occasional floods with overflow of the river banks, and/or rising of the level of the groundwater to the surface, limits the choice of crops. A separate problem is the quality of the floodwater and/or that used for irrigation, i.e., salt and pollution content.

Therefore, ameliorative treatments have to be oriented towards flood control, and lowering groundwater level in skeletal and sandy textured fluvisols, to achieve stable and high yields; irrigation is also required. A good supply of water can be achieved by regulating the depth of groundwater (if it is in connection with the river water level), by adjusting dam levels and maintaining an optimal level of groundwater and capillary supply to the rhizosphere of crops.

6.3.2 Class II: Semigleyic Soils of A-...-G_r Sequence of Soil Horizon

This class consists of systematized soils in which pedogenesis in the upper part of the profile occurs predominantly as an automorphic process, and at depths greater than 100 cm there is a gley horizon under the influence of groundwater. In Croatian soil science, such a type of hydromorphism is named meadow or semigley hydromorphism. The research of Bogunović et al. (1996, 2001) found that a number of automorphic soils can be called meadow in relation to the type of wetting—with a groundwater gley horizon below 100 cm depth.

6.3.2.1 Semigley

The name of the soil type. Its name relates to the specific water regime because the contemporary name semigley means "half-gley" (semi—half, partial), because the gleyization is only partial—the gley horizon occurs deep in the C horizon—the parent substrate. This water has a positive influence on the rhizosphere. This soil type is described as meadow soil in Croatian literature, and this name is given according to the natural meadow vegetation that occurs on it. The name of this soil is similar in other languages.

Genesis and sequence of horizons. They are representative of the conditions of genesis. In the upper layer, where automorphic soils are dominant, the following sequence of horizons occurs: A_{mo}–C–Gr with terrestrial material, usually a molic type of humus-accumulative horizon, below which occurs the gleyless parent substrate—C, and inside this at a depth below 100 cm a groundwater gley horizon.

Distribution. This soil type covers an area of 89,901 ha or 1.6 % of Croatian land (Fig. 6.22). It occurs throughout the country, particularly in the valleys of major rivers in the Pannonian region in the lower loess terraces.

It is present in the middle and upper Podravina, as the dominant unit in Posavina in association with other hydromorphic soils, and in all regions and subregions of the whole country it occurs as inclusions in river valleys. As a rule in the cross-section of flooded river valleys, we find fluvisol as recent deposits in the coastal zone, as semigley in

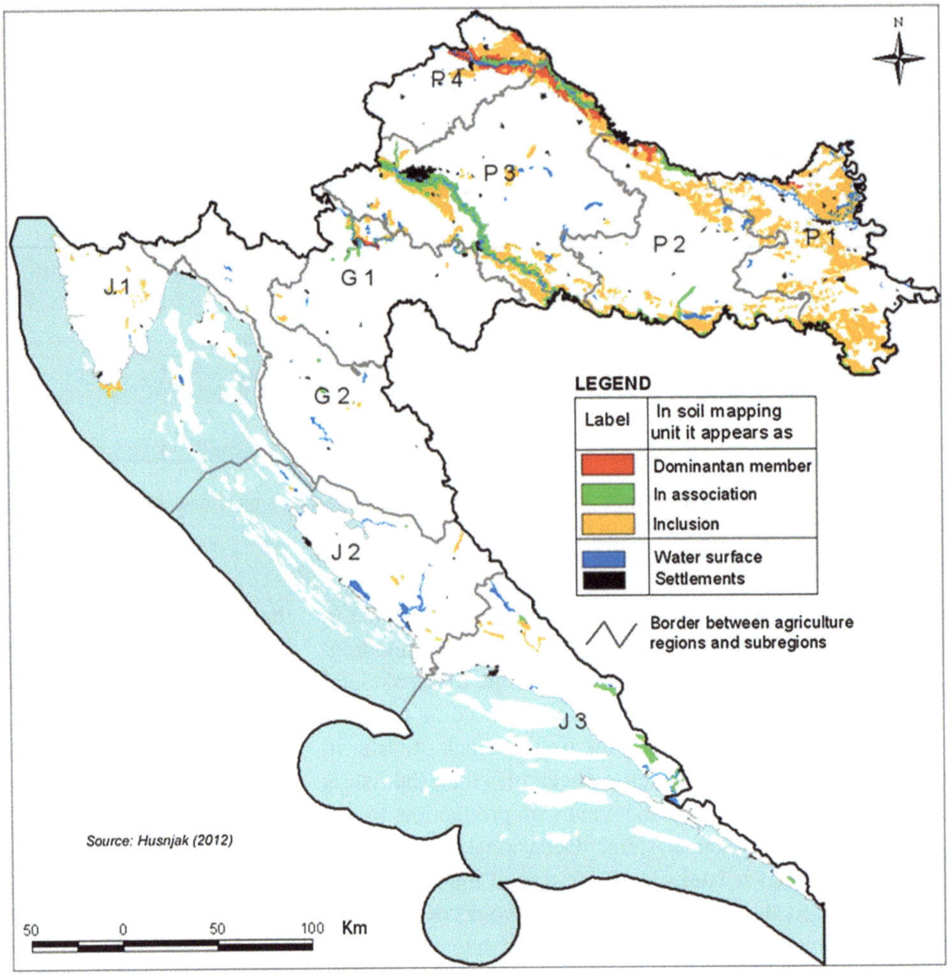

Fig. 6.22 Distribution of semigley in the pedosphere of Croatia

the central zone, and peripherally, on lower slopes of gley soils. In terms of pedogenesis, there are a number of combinations and transitional forms among these soils.

Semigley has a high content and quality of mollic humus, the chemistry of non-calcareous and low calcareous varieties is neutral or weakly acid, but in alkalized varieties it is strongly alkaline, and the pH generally increases with depth. The soil has high CEC and high base saturation (80–90 %), Ca^{++} and Mg^{++} ions are dominant but in alkaline varieties Na^{+} occurs, which disrupts the very favorable physical features. Semigley is, as a rule, well supplied with plant-available nutrients, especially

Photo 6.57 Semigley soils are valuable agricultural soils with excellent soil fertility—Virje in central Podravina (P-3)

potassium, but is somewhat less well supplied with phosphorus and nitrogen.

Land use. Semigley has various uses. As one of our most fertile soils, its high potential and effective fertility is due to the large depth of the active profile and humus horizon, good air–water and thermic conditions, and the particularly affordable supply of water (Photo 6.57).

If used as a natural meadow it has high quality and very favorable botanical composition, high participation of legumes (*Trifolium* sp.) and graminea (*Arhenatheretum elatioris*), and gives high yields of hay of excellent quality. As arable land, it can be used for growing a wide range of crops—wheat, sugar beet, rapeseed, sunflower, and corn; all provide high and stable yields. Semigley is a very favorable soil for the growing of sugar beet as a high-demand crop.

Basic measures of amelioration include flood protection and control of groundwater level.

6.3.3 Class III: Hypogleyic Soils of A–G_r Soil Horizon Sequence

This class of phenomena is characterized by soils with high groundwater level at a depth of less than 100 cm, which is retained longer in the winter and autumn periods, and in a dry summer the level significantly declines. There are two types of soil in this class.

6.3.3.1 Hypogley

The name of the soil type. The name derives from the Greek *hypo*—below, low, which indicates that the origin of hydromorphism is groundwater and that the gley horizon is positioned in the lower part of the profile.

Genesis and sequence of horizons. This is well described by the name, i.e., this is a hydromorphic soil that predominantly occurs within the hypogleyic regime of

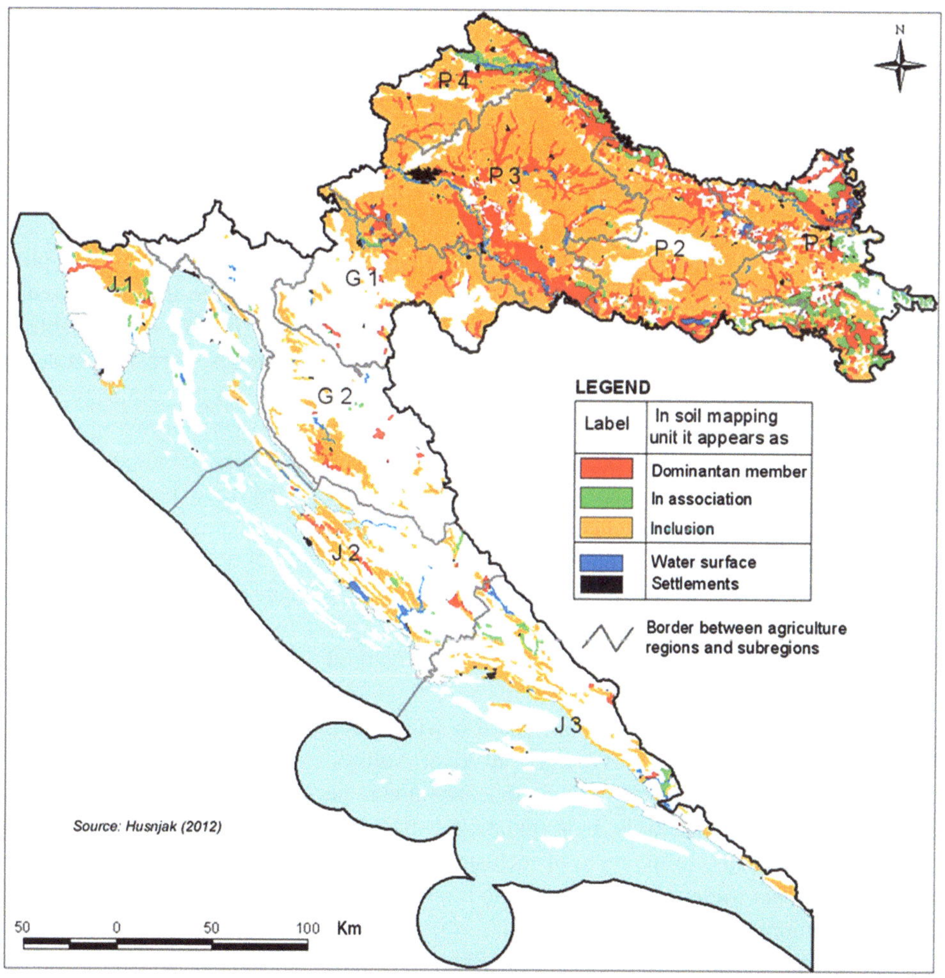

Fig. 6.23 Distribution of hypogley in the pedosphere of Croatia

Photo 6.58 Textural heavy clayey hypogley of Upper Posavina used as arable land and oil-exploitation field—Žutica near Ivanić Grad (photo: Kisić)

hydromorphism, dominated by the impact of high groundwater levels over a greater part of the year. The G_r horizon occurs at a depth of less than 100 cm. The profile of this soil consists of deposited river sediments, the material is located relatively close to the riverbed, and groundwater is at a depth of less than 100 cm at which occurs G_r—the reductive gley horizon with its typical blue-green color. If the groundwater level varies within the wet zone, there are fluctuations between exchange and oxidation processes that give rise to the characteristic G_{ox} horizon. In virtually all occurrences of this soil there is no soil leaching, because the dry part of the year is dominated by ascending flows of water back to the rhizospheric zone. So, eventually, leached bases return to the upper part of the soil.

The WRB name. Gley soils

Distribution. Hypogley is found in all agricultural regions (Fig. 6.23).

As can be seen, as a dominant soil type it is widespread in the Carpathian region, in virtually all river valleys, including the valley of the P-1 subregion and along the upper part of the Drava River in association with other, genetically close soils. In the form of inclusions, it can be found in narrower valleys, mainly in the P-1 subregion in western Syrmium.

Physical and chemical properties. They are very different and dependent on soil texture. More favorable is hypogley of light loamy or sandy texture. The fertility of this soil type depends on the texture and depth of the water table—groundwater level during the growing season. Therefore, if it is light, loamy and sandy, it is generally more fertile than if it is heavy—clay soil.

Land use. Hypogley soil can be used as arable land or good meadow if it is not vertic (Photo 6.58). Dependent on texture it can be used as a natural habitat of meadows and forests of valuable oak with the lighter textured varieties used as arable land. It requires hydroamelioration—pipe drainage and complex practices oriented towards improving the chemical properties of the soil. Heavily textured hypogleys do not necessarily respond well to expensive drainage operations and are best left under natural vegetation.

If hypogley is used as arable land it is recommended to carry out the traditional practice of autumn–winter plowing to expose the soil to winter frost, which forms a favorable soil structure.

6.3.3.2 Humogley

The name of the soil type. The name originates from the word humus—organic matter and gley—marsh mud layer. In Croatian literature it has been described as dark marsh soil, hydrogenized humus soil, or wetland (boggy) soil (Photo 6.59).

Genesis and sequence of horizons. Humogley lies "between" semigley and hypogley soil. Humogley occurs at sites with a so-called boggy type of hydromorphism, characterized by large fluctuations of the water table—in one part (winter/autumn) of the year the water table rises and groundwater rises to the surface, while in the other part it falls

Photo 6.59 Humogley in Klisa near Vukovar. After heavy rains soil crusts can form on bare soil

to a few meters below the surface. The wetland is covered by very luxuriant wetland vegetation, predominantly reed (*Pragmites* sp.). The profile of this soil has a humus-accumulative horizon of mollic type with signs of hydromorphic conditions—A_{mo}, a black color and depth greater than 50 cm. Below this is a secondary oxidized gley horizon—G_{so} to a depth of 100 cm or more. It forms under conditions of arid or semiarid climate in depressions or flattened relief forms (terraces and loess plateau) that are influenced by boggy hydromorphism. Its genesis, compared to semigley, is distinguished by a more intensive hydromorphism.

The WRB name. Gleysols

Position in an evolutionary series. This depends on the wetland regime—the regime of hydromorphism. If humogley remains in the same regime of hydromorphism, evolution is directed towards peat soil—histosol formation. If hydroameliorative practices fix the water table, so it never rises to the surface of the soil, then the evolution is directed towards automorphic soil—rendzina on marshy deposits.

Distribution. Humogley covers an area of 64,555 ha or 1.2 % of Croatian land territory, predominantly in the Pannonian region (Fig. 6.24).

It occurs mainly in the Holocene terrace of the Danube and Sava rivers in the eastern Pannonian—P-1 subregion, especially in Baranja, the entire stretch of the Danube, then on loess terace between Osijek and Vukovar, Klis and Bobota, near Lipovac and Drenovci, and along the rivers Sava, Bosut, and Spacva in the extreme south-east of the Pannonian region. There is also a protected area—Kopački rit, with an extraordinary natural wealth of wildlife, especially birds, fish, and wetland vegetation.

Physical and chemical properties. The properties of humogley depend on the texture of the soil. Vertic, especially saline and alkaline, soil has an adverse physical characteristic; it is plastic, gluey-adhesive soil in wet condition, having a high water-holding capacity but small quantities of plant-available water in the soil. The period of moisture of the soil which is convenient for tillage, is very short. Humogley with lower clay content due to the high humus content and its high quality, has more favorable physical properties. All humogley soils have favorable chemical properties such as high humus content, adsorption capacity, base saturation, and nutrient supply. Overall, humogley of loamy texture such as we find in Baranja is a soil of very high soil fertility; it is less fertile in vertic, saline, and alkaline varieties. For stable and high yields and a wider variety of crops, hydro- and agromelioration should be carried out on these soils, as a prerequisite for growing cultivated plants. The high fertility potential of humogley comes to the fore only after hydroamelioration—flood protection and lowering of the water table. In nature, humogley occurs in association with other hydromorphic soils (fluvisol, semigley). Therefore, when one considers reclamation of humogley one must take into account the effect of lowering the water table on the other soils. Lowering of groundwater levels is problematic for semigleys that require (particularly in lightly textured, sandy or skeletal variants) irrigation in the summer, which is the critical period of crop growing for maintaining stable and high yields. The direction and intensity of soil improvement by hydroamelioration in this case is decided based upon the proportional representation of the individual soil types and, of course, the economics of this practice. Humogley responds well to fertilization and mineral fertilizers, but the vertic variety of humogley requires irrigation and deep tillage.

After the performance of such ameliorative soil improvement only intensive cultivation of field and vegetable crops can be considered, with high and stable yields. The reclaimed area is used for growing arable crops—wheat, corn, sunflowers, sugar beet, alfalfa. One expects high yields of all crops, particularly sugar beet and corn.

Without amelioration the farming operation is unstable and insecure. No-tillage farming is a recommended practice on humogleys, especially in the cultivation of small grains and corn as the dominant crops in the field. As the results of our analysis show, the entire area of the P-1 subregion is affected by desertification processes with a low annual quantity and/or irregular distribution of precipitation (Bašić and Bašić 2007). For example, in 2012 in the eastern Pannonian subregion in the spring the soils were under

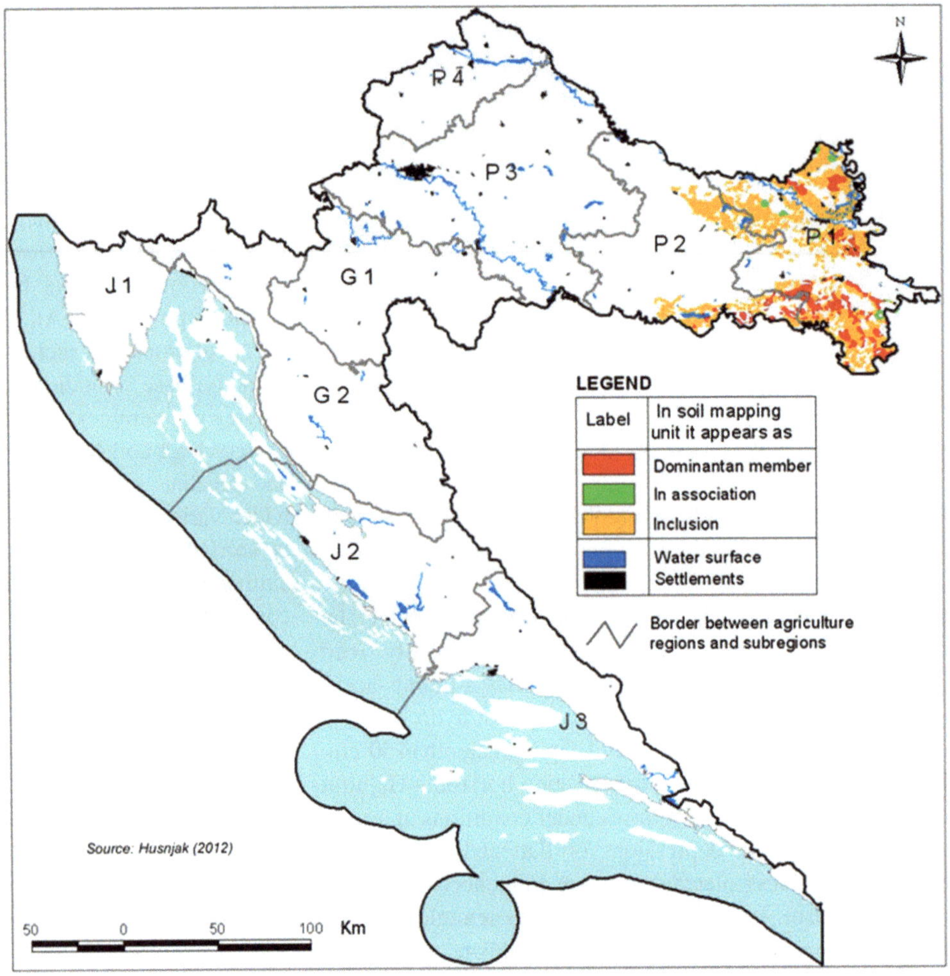

Fig. 6.24 Distribution of humogley in the pedosphere of Croatia

completely dry conditions. The last serious rain fell seven months before.

6.3.4 Class IV: Amphygleyic Soils of A_a–G_r–G_{so}–G_r Soil Horizon Sequence

The fundamental characteristic of this class of hydromorphic soils is its extremely swampy conditions. The name "gley" is found in all soil types of this class. This Russian word means muddy soil mass, and in soil science it was introduced as an indicator of soil horizons dominated by reduction processes, caused by the influence of saturation of some horizons by water. The processes of reduction under conditions of full saturation with water are called gleyization. In Croatian pedological literature, this soil is described as mineral-marshy, hydrogenized or gleyed wetland soil. Eugley or "the right" gley (wetland) soil is defined as soil which to a depth of 100 cm has clearly differentiated G_{so} and G_r subhorizons, resulting from stagnation of the upper (epi) water (from flood or lateral inflow from slopes), lower or groundwater (hypo), or both (amphy) waters in the profile. In the taxonomy of Husnjak (2012) the amphygley class is classified into three types of soil.

Distribution. As epigley and amphygley in previous taxonomies were defined as one soil type, eugley, the total area covered by both is 499,526 ha or 9.0 % of Croatian land, mostly occurring in the Pannonian subregion. The distribution of certain types of soil of this class has yet to be determined.

6.3.4.1 Epigley

The name of the soil type. The name is a compound of the Greek word epi—upper, which means gleyization by the upper water and the already-described term gley, and so it could be called "upper gley."

Genesis and sequence of horizons. This is a consequence of the epigleyic regime of hydromorphism, which forms the soil profile of epigley: an A_a–G_r–G_{so} sequence of horizons indicating the processes that occur in this soil, clearly visible on Photo 6.60. It occurs in slight topographic depressions in the flood zone of all Croatian rivers. The humic horizon shows the character of humus under anaerobic conditions and below it follows the G_r subhorizon with

Photo 6.60 Cracks in vertic non-calcareous epigley

dominant processes of reduction and diffuse transition to the G_{so} subhorizon in which oxidized forms of iron remain.

Sufficient amounts of surface water, flooding from watercourses, or lateral inflow from surrounding slopes can lead to water retention that lasts until removal of the water by evaporation and transpiration processes. Obviously, lush wetland vegetation is naturally present on this soil type.

The WRB name. Gleysols

Position in an evolutionary series. This depends on the hydromorphic regime, which is influenced by the building of roads and infrastructure for oil and gas exploitation, underground pipes for oil transport, and the digging of canals for flood protection and amelioration. All of these anthropogenic influences affect the hydromorphic regime in all areas.

Physical and chemical properties. The properties of this soil are dependent on texture. The soil has a heavy texture with high montmorillonite content—a type of clay that leads to several consequences. In the dry state it contracts and in the wet state it expands, leading to a volume increase causing pedoturbation and a visible "bump" in the land. Chemical features are favorable: a high CEC, absence of acidification, high humus content and available nutrients.

Guidelines for management. Epigley is unsuitable for tillage in its dry condition when it has very strong resistance to any tillage treatment, and under wet conditions, it is plastic, very sticky and can spread, causing the wheels of machinery to slip. Deep plowing in autumn and the formation of an optimal structure after exposition to winter frost is a recommended practice.

Land use. As natural pastures of adverse botanical composition or as arable land that is not easy to manage, since it requires great expertise and opportunities for customization.

Photo 6.61 Pseudogley–gley on Pleistocene loams near Jastrebarsko in the forest—pedunculate oak and common hornbeam (*Carpino betuli* and *Quercetum roboris*). Photo: Vrbek

6.3.4.2 Pseudogley–Gley

The name of the soil type. The name is a compound of the type of soil—pseudogley, thus soil wetting with excess rainwater, and gley, which indicates that pseudogley is exposed to gleyization by underground water up to 100 cm of depth.

Genesis and sequence of horizons. This is described in the previous text (Photo 6.61). The sequence of horizons in a profile is $A_{a/p}$–E_g–S–G_r, while under forest vegetation on the surface there may occur an $O_{l/f}$ horizon. Pseudogleyization is usually a primary process that is a consequence of the stratigraphy of the substrate. It occurs most often in river valleys on the foot of slopes with gley soil. Naturally, it is overgrown by swampy vegetation of low economic value.

The WRB name. Stagnosols, gleyic

Distribution. The distribution of pseudogley–gley is relatively small and amounts to 84,713 ha, which means 1.5 % of the land area, mostly in the valleys of the Sava and Drava River, where it is the last habitat of valuable forests of oak and hornbeam (*Querco-Carpinetum* sp.) (Fig. 6.25).

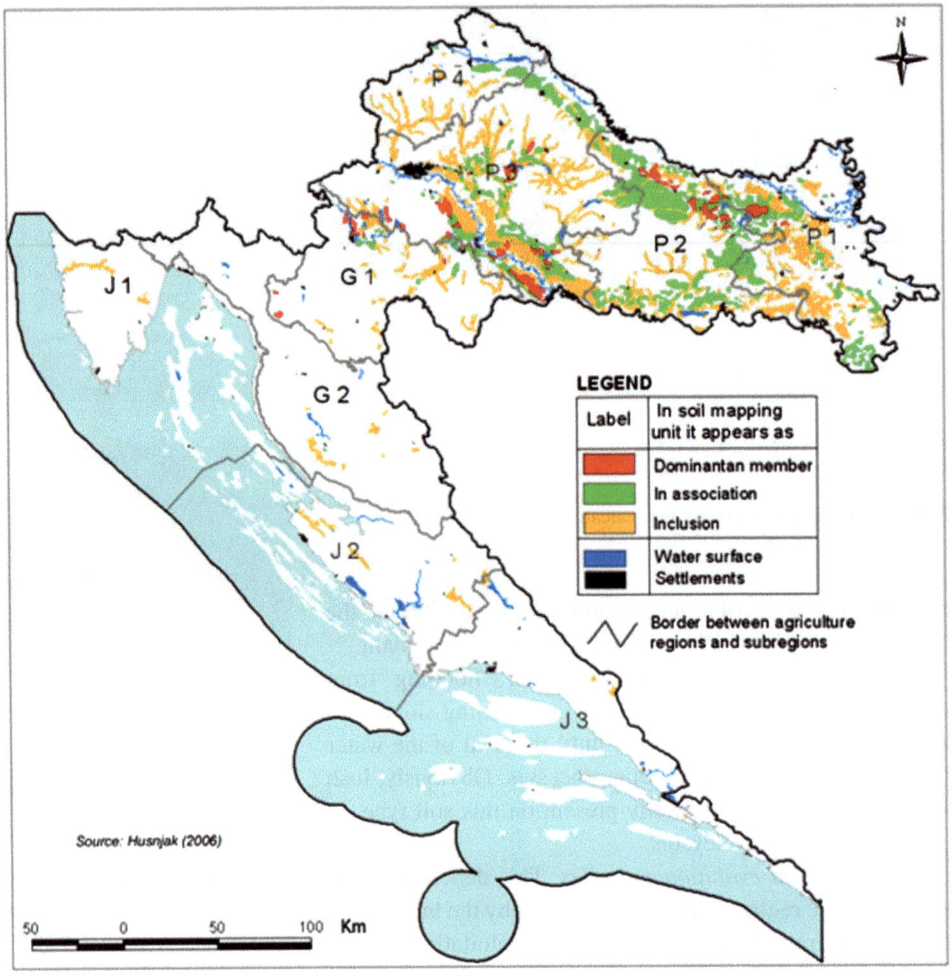

Fig. 6.25 Distribution of pseudogley–gley in the pedosphere of Croatia

As the dominant pedosystematic unit in the cartographic unit pseudogley–gley is widespread in the lower and upper Podravina and Posavina, more lightly textured in Podravina but heavier in Posavina. The building of a basic network of canals for drainage of agricultural soils and the construction of a road network that monitors oil and gas sites in these areas have significantly disrupted the water regime. Thus, the stability of the lowland forest ecosystems on the soil has become disturbed, and consequently we are witnessing a process of death in these most valuable forest species.

Physical and chemical properties. The properties of this soil type depend on the character of the process of pseudogleyization (primary or secondary) and the characteristics of the substrate on which it arises. The substrate is mainly leached loess or sediment of fluviative and/or colluvial origin, on positions of lower relief, and Pleistocene noncalcareous clay. The common feature is the high content of silt, which indicates its origin as an eolian material. Vrbek et al. (2003, 2009, 2011) studied soils of a lowland swamp forest and found textural differentiation and a higher clay fraction content in the deep gley horizons. The chemistry of these soils is very acid, the pH in $CaCl_2$ is below pH 4, which means that in the soil solution there are indications of Al^{+++} ions, which are phytotoxic to some plants. The CEC is low. Supplies of plant-available phosphorus and potassium are low and medium, respectively.

Land use. This soil is a natural habitat of good sessile oak and hornbeam forest or marsh meadows dominated by *Carex* and *Juncus* species of too low a value for any use. Vrbek (2009) states that hornbeam is an indicator of the regime and intensity of hydromorphism, because it withstands temporary short-term flooding, but cannot withstand stagnant and high groundwater, and therefore hornbeam grows best where the groundwater is at a depth of 2–3 m from the surface.

Guidelines for management. Management of this soil is very complex. The best form of use is as a good forest soil. This soil is unfavorable for agricultural use without intervention by the hydroameliorative practices of surface drainage and water table lowering—regulation of groundwater level. These are followed by agromeliorative procedures that create a deeper active top layer and improve the chemical characteristics within this layer, opening the door to sustainable arable farming. Ameliorative interventions are complex, and in relation to the reclamation of other types of soil they are considerably more expensive. However, long-term project feasibility even after the performance of these complex measures is questionable as certain features restrictive to the growth of particular plants can be mitigated by reclamation, but not eliminated forever.

6.3.4.3 Amphygley

The name of the soil type. It derives from the Latin word *amphy*—both, and describes a soil with gleyization caused by both types of water saturation: underground (hypo) and surface water (epi), meaning two G_r horizons, between which is a less gleyed, secondary oxidized G_{so} subhorizon.

The WRB name. Gleysols

Genesis and sequence of horizons. This is related to the specific topographical and hydrological conditions. In the water-saturated zone, a gleyic reduction subhorizon is formed—G_r. This leads to reduction of Fe^{+++} to Fe^{++}, Mn^{+++} to Mn^{++}, sulfate to sulfide, nitrate to ammonia or gaseous nitrogen, and reduction of organic compound derived gases: ammonia, hydrogen sulfide, methane, phosphine, etc.

The reduced compounds give the G_r subhorizon a greenish, gray, or bluish color (Photo 6.62). The high content of clay reduces porosity and permeability to water, increases adhesion and plasticity, and corrupts the physical properties of the soil.

In the zone of periodic water saturation, a secondary oxidized G_{so} subhorizon occurs. During the time it is saturated, reduction processes dominate but after withdrawal of the water oxidation processes occur which leave a noticeable rusty-reddish colored horizon, generated by oxidation of iron following a fall in the groundwater level. Under these conditions there is a typical hydrophilic, and in the case of acidity an acidophilic meadow vegetation, whose representatives are *Deshampsia cespitosa*, *Carex* sp., *Luzula* sp., and *Juncus* sp. (Photo 6.63).

The residues of plants are humificated under anaerobic conditions, creating a special A_a horizon, with clear signs of the beginnings of hydromorphism or peat—a T horizon.

Distribution. This soil occurs in all major river valleys in association with other hydromorphic soils and in lower topographic positions (depressions) prone to flooding and a high water table. These are then wetland conditions with special flora and fauna, protected according to the Ramsar Convention. In a valley of a river basin is the famous Nature Park Lonjsko polje, in which one can find an important Cigoc village—the first European stork village (Photo 6.64).

Physical and chemical properties. The properties of amphygley (Table 6.32) are very heterogeneous. Soil texture varies from light sandy to heavy, clayey, vertic, which facilitates favorable or unfavorable physical features.

Table 6.32 shows some of the physical and chemical properties of amphygley of the Pannonian subregion.

Because of the high content of the montmorillonite type of clay, vertic amphygley is a cold, slightly porous, poorly permeable, very sticky and plastic soil, with a short period of moisture favorable for tillage treatment (the "minute" soil). A transition between the wet state and the dry one changes the volume, causing pedoturbation and a visible "bump." The chemical characteristics also vary depending on the features of the layers that form the amphygley. There are non-calcareous and calcareous amphygleys. The chemistry is, on average, almost neutral, but also varies in the range from pH 5.6 to 7.2 in the top horizon, and going deeper it can reach pH 8. The humus content in the A_a horizon varies from poor to very rich: 1.4 to 9.1 % humus. Supply of plant-available nutrients varies from very poor to very rich. It is important to emphasize that under anaerobic

Photo 6.62 Amphygley vertic in the valley of the Bednja River near Novi Marof (P-4)

Photo 6.63 Indicator species in meadows on amphygley—*D. caespitosa*, *Juncus* sp., *Carex* sp

Photo 6.64 *Left* Storks enjoy the first village in Europe that was colonised them, building their nests that weigh several hundred kilograms. *Right* A typical landscape in which amphygley occurs—plenty of cattle and countless birds in gardens and in the wild, and in the swamps plenty of food for storks

Table 6.32 Some physical and chemical properties of amphygley of the Pannonian region

Soil horizon	pH in nKCl			mg P_2O_5/100 g of soil			mg K_2O/100 g of soil			Humus (%)			Clay (%)		
	Average	Minimum	Maximum	Average	Minimum	Maximum	Average	Minimum	Maximum	Average	Minimum	Maximum	Average	Minimum	Maximum
A	6.5	5.6	7.2	8.7	1	40	10.6	4	31	3.3	1.4	9.1	36.9	22.3	54.4
G_r	6.9	5.4	7.6	3.8	0	35	7.5	4	21.5	1.3	0.2	3.6	36.6	23.7	50.1
G_{so}	7.5	6.1	8.6	–	–	–	–	–	–	–	–	–	25.3	12.8	41.3

Source Bašić et al. (2003)

conditions amphygley phosphorus is fixed in insoluble ferrophosphates, and potassium is well supplied. The total content of nitrogen is high, but little is in a plant-available form. There is a risk of denitrification—nitrate reduction to gaseous nitrogen—its loss from the soil. The CEC is high in clayey and humus-rich amphygley, and base saturation is high. If salt content in the water is high, it can lead to salinization and alkalization.

Photo 6.65 Truffle (*Tuber melanosporum*) is a highly prized fungus that develops its underground tuber on the root of oak (*Quercus robur*). A famous site is the oak forests on lighter textured, calcareous amphygley in the Mirna valley beneath the historic Istrian town of Motovun

Photo 6.66 *Left* Wetland meadow on vertic amphygley—Kukunjevačko polje by Lipik. *Right* Dying pedunculate oak forest of Žutica near Ivanić Grad

Photo 6.67 Drought after soybean emergence on the vertic amphygley of farm Poljodar-Daruvar (P 3). Cracks in the soil allow the separation and extraction of sharp hard lumps. The cracks damage the roots of soybean. A high level of professional knowledge is necessary for management under such conditions

Land use. The natural vegetation that occurs on all amphygley soils in Croatia is marshy lowland forests, among which occur both economically and ecologically important forests of oak and hornbeam (*Carpino betuli*, *Qercetum roboris*) including a number of very famous and protected forests: Spacva in the P-1 subregion, jungles Prašnik and Muški bunar in the P-2 subregion to Motovun forest in Istria (P-1), known for its exceptional natural heritage: it is the site of a rare and highly prized culinary delicacy—truffle mushrooms (Photo 6.65).

Many localities in Posavina, the natural sites of vertic soils of Lonja polje and in Podravina, are influenced by many interventions, like the excavation of channels for the primary reception of water from agricultural land, the creation of a series of hydroelectric plants on the Drava River, and construction of road infrastructure in Posavina, leading to disturbance of the water regimes. These interventions, compounded by the sporadic attacks of pests and diseases, cause drastic changes in the quality of groundwater (contamination and increase in temperature of the Sava River water because of its use for cooling the nuclear power plant) leading to the sporadic death of oak on amphygley soil (Photo 6.66).

In terms of its use in agriculture, meadows on vertic varieties of amphygley are unfavorable with a botanical composition of low value, high yields of hay of low quality. This situation is a little better on lighter textured varieties. Amphygley has a high potential and low real effective fertility.

The main limiting factor for production in this soil is water surplus. The amelioration practices on this soil are oriented first towards flood elimination, followed by drainage of surface water, and lowering of the underground water table. These procedures should be amended by hydro- and agrotechnical amelioration—drainage by channels or pipe, deep loosening, humification of plowed plant residues, and then all operations that affect the repair of physical properties of soil.

Such complex interventions can improve amphygley to an extent that ensures stable and high-yield arable crop production. It should be emphasized that the amelioration of vertic amphygley is a very expensive operation, and management of this soil and maintenance of ameliorative systems requires a high sensibility and expert knowledge (Photo 6.67).

If you are unable to provide the challenging level of management of these soils, we recommend that it is better to leave it to natural use—afforestation or pastures to fertilize and maintain.

6.3.5 Class V: Peat Soils of T–G_r Soil Horizon Sequence

6.3.5.1 Histosol

The name of the soil type. Derives from the Greek word *histos*—tissue, which suggests that in the peat horizon (T) one recognizes the tissue of the plants from which peat is formed.

Genesis and sequence of horizons. This is connected with the wetland condition. The main process of genesis is formation of an endomorphologically recognizable and marked peaty—T horizon, the depth of which exceeds 30 cm. It was formed by humification under anaerobic wetland water conditions. There is high, transient, and low

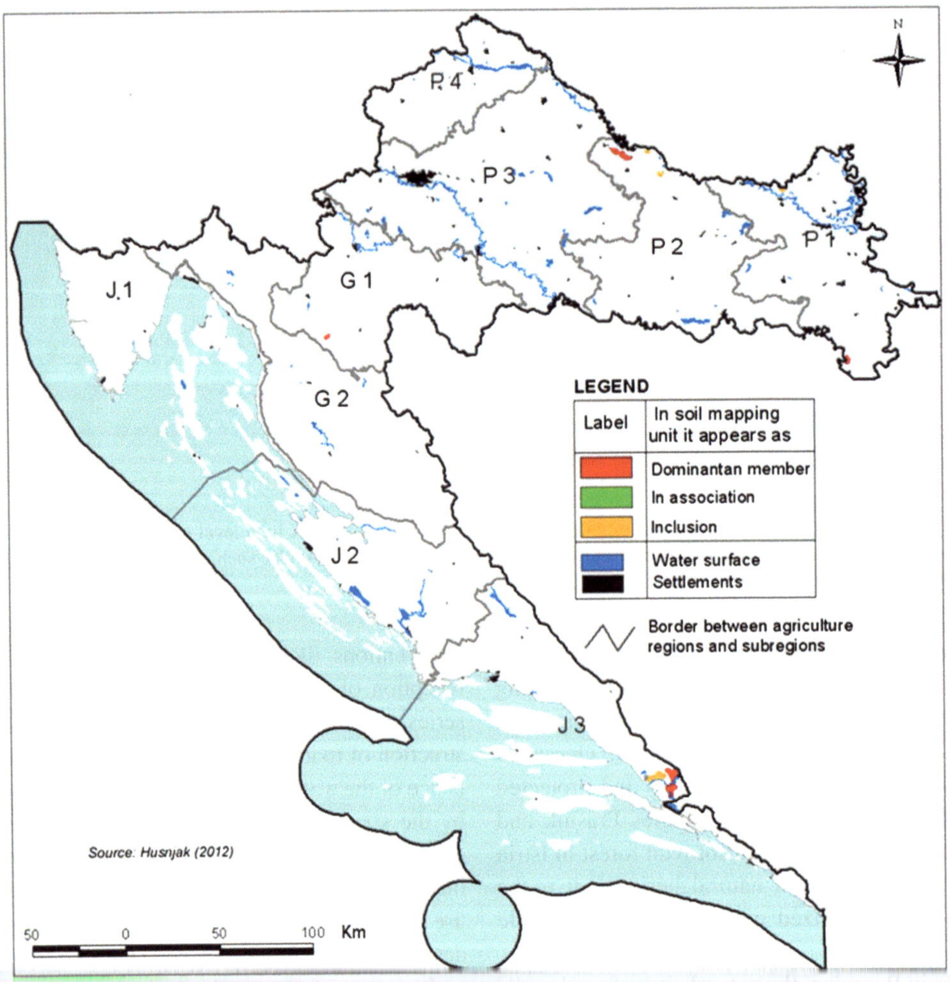

Fig. 6.26 Distribution of histosol in the pedosphere of Croatia

Photo 6.68 Alder forests Crni jarki (black rills)—Kalinovac with a very rich and dense vegetation (*left*) and high production of organic matter per unit area. *Right* The peat bog "Beričanjki"—Kalinovac (P-3)

peat. In Croatia low peat dominates as a topogenic creation generated in plains on low topographic positions where there is excess moisture during all or most of the year. It is formed by incomplete and slow decomposition under anaerobic conditions of the residue of dense wetland vegetation that accumulates on the surface of the wetlands. For the formation of the T horizon, several thousands of years are necessary, with gradual accumulation of peat from the top of the mineral soil upward.

WRB name: Histosols

Distribution. The distribution of histosol is practically symbolic; only 2,577 ha of land or 0.05 % of Croatian territory (Fig. 6.26).

A prerequisite for the formation of peat is a large amount of organic matter. This may be created "in situ" through the lush wetland vegetation or via the plant residues of deciduous trees carried by flowing water and concentrated in specific localities, i.e., old river meanders or depressions, where it sediments and gradually undergoes humification, which occurs to a greater extent under anaerobic conditions, creating the peat's T horizon. We observe this process in the Podravina swampy alder forests, first as a peaty process in epi- or amphygley soils. In late autumn a mass of leaves fall onto the water and through the action of wind this mass is transported over the water's surface to a point protected from wind where it accumulates and slowly generates peat. This is repeated annually and the peat layer gradually "grows." On the surface of the organic material we can identify plant species, but with increasing depth and fermentation this is no longer possible and finally we reach a deep black organic substance which is in the advanced stage of humification. Photo 6.68 shows the famous "bright black" alder forests of Podravina, and not far from the forest there is a peat bog named "Beričanjki", seen in the right part of the photograph.

The Drava was once much larger, as was the peatland. The old town of Djurdjevac was erected on a "hill" in the middle of the peatland which became an insuperable obstacle to the penetration of invaders.

Land use. The peat can be used for various purposes, including preparation of a substrate for greenhouses and for growing potted plants. Large areas of histosol have been drained, the water table lowered, and the soil cultivated. One of the largest peatlands and one of the best and most fertile agricultural areas is the delta mouth of Neretva River

Photo 6.69 "Everlasting" peat fires in the Livno area of our neighboring country, Bosnia & Herzegovina, are a common occurrence, but also an insoluble problem for the local people

Photo 6.70 Histosol is a very fertile soil for growing vegetables and flowers, especially root, tuberous and bulbous varieties

(A-3 subregion). Very turbulent mineralization of the humus in peat follows cultivation of histosol. This releases significant thermal energy and methane gas, which can cause spontaneous combustion of the peat. Such peat fires are usually very difficult to extinguish because the smoke usually emerges at some distance from the actual fire (Photo 6.69). The fire is virtually impossible to locate. Current climate changes are causing more frequent fires of this type globally.

In the Busetina area, near the city of Virovitica in Central Podravina (P-2), the area of Bušetinski Berek was successfully ameliorated and cultivated following a study and project created by Butorac (1983).

In the cultivation of histosol the first practices include protection from flood water, and lowering and fixation of the water table so that it does not penetrate the peat layer. Mineralization in the deep layer of peat after several years of exploitation has consequences for the settlement of the soil surface, because it reduces the volume of the surface layer. This produces the requirement of constant leveling of the soil surface and reconstruction of drainage networks, as this reduces the volume of subsidence.

Peat soils are very suitable for growing sugar beet, vegetables and flowers, especially root, tuberous and bulbous vegetables and flowers (azaleas) if further conditions are favorable. For this purpose it is necessary to provide irrigation (Photo 6.70).

6.3.6 Class VI: Anthrosol Hydromorphic of P-...-$G_{r/so}$ Soil Horizon Sequence

6.3.6.1 Rigosol

The name of the soil type. The name was introduced into soil taxonomy by analogies with terrestrial and semiterrestrial anthropogenic soils. It is therefore a hydromorphic soil in which at least two genetic soil horizons are mixed by deep plowing; this plowing creates a cultural soil as a substrate for plant growth. The name is derived from plowing—anthropogenic plowing, meaning a mixing of several soil genetic horizons with particular hydromorphic features into a soil with new characteristics, which are more favorable for the breeding of cultivated plants.

Endomorphology—the sequence of horizons in the profile—is not uniform and depends on the original soil type which was anthropogenized by amelioration and trenching, and these are usually fluvisol, humogley, epi- and hypogley or histosol. For cultivation one first needs to protect the soil from "foreign waters," i.e., control of flood waters or regulation of the groundwater level. This means radically changing the conditions of pedogenesis in a new direction characteristic of terrestrial–automorphic soils (Photo 6.71).

This soil type is widespread throughout the country. In each region, at least part of the subregion, hydromorphic soils are cultivated with more or less changes to the natural characteristics of the soils. In the series of interventions plowing is obligatory.

Photo 6.71 *Left* At the mouth of the Neretva River rigosol occurs in association with other drained hydromorphic soils. *Right* Powerful trenching plow

Table 6.33 Taxonomy of rigosol hydromorphic soil

Name of taxon and base of classification	
Variety (based on organic matter content)	Form (based on $CaCO_3$ content)
For all subtypes • Moderate content 30–50 % • Rich content 50–75 % • Very rich content > 75 %	For all varieties • Calcareous • Non-calcareous

These soils follow a terrestrial genesis and evolutionary path. First, there is transformation of the aquatic A_a to a mollic A_{mo} horizon, showing gradual loss of morphological features through the reduction process; this is followed by a leaching process and the corresponding climax stage as in terrestrial soils in surrounding areas. There follows the formation of a $(B)_o$ horizon reaching a cambic, then luvic and/or pseudogleyic stage. Taxonomy of the unit lower than subtype is made according to the criteria shown in Table 6.33 (Photo 6.72).

If this soil is lightly textured and calcareous it may be a very fertile and valuable soil, especially for vegetable growing.

Photo 6.72 Using flood prevention and regulation of groundwater level in the hypogley peaty soil, with trenching/plowing and mixing of the peaty A_a horizon with mineral material in Višnjica, near Slatina in the P-2 subregion, a very fertile anthropogenic soil has been produced. In an excessively wet year the water table is quite deep

6.3.6.2 Drained Soil

The name of the soil type. The name specifies the type of anthropogenic intervention, which obligatorily includes the installation of perforated plastic drainage pipes at a proper depth in the soil.

Genesis and sequence of horizons. This depends on the type of soil that is drained, but on the surface is a P horizon, which may contain features of the genesis under hydromorphic conditions or have the characteristics of a T horizon. There follows a sharp transition to the next horizon—G_{ox}, G_r, or C.

The WRB name. Anthrosol

Position in an evolutionary series. This depends on many factors, first on the maintenance of the drainage systems which should allow maximum aeration of the soil, oxidation and intensification of aerobic microbial processes. Performance of regular deep loosening belongs to a group of interventions aimed at maintenance of the ameliorative system. Regulation of hydromorphic soils brings with it quite rapid change in the features of the humus in the A horizon, especially those indicating its origin and aquatic features. In addition, there is strong mineralization of organic matter, especially in peat soils, with emission of huge quantities of CO_2—a greenhouse gas. One should recall that hydromorphic soils are a plentiful source of methane emissions, which is a "very effective" greenhouse gas (Table 6.34).

Distribution. The distribution of this soil type is not known exactly. There is a total area of 163,000 ha of drained soils, or 2.9 % of the land area of Croatia, but this includes some areas of semiterrestrial drained soils—pseudogley. Unfortunately, over parts of this area maintenance is lacking, leading to clogged drainage pipes and decay of the system. Through privatization, part of the soil came into the ownership of totally unqualified people, and in the channels, the collectors of water from drainage pipes are completely covered by mud, obstructing normal water flow. The alders are growing and are ready for felling. This is a terrible outcome of the dealings by our decision makers regarding an extensive national treasure (Fig. 6.27).

Table 6.34 Taxonomy of drained soil

Name of taxon and base of classification	
Variety (based on depth of drain pipes)	Form (based on texture)
For all subtypes • Shallow <50 cm • Moderately deep 50–75 cm • Deep 75–100 cm • Very deep 100–125 cm	For all varieties • Sandy • Loamy • Clayey

Prior to installation of drainage measures a pedological study must be carried out to detect the origin and intensity of hydromorphism and to provide justification for the installation of a drainage scheme. This is followed by a design study for the drainage pipe system (depth of

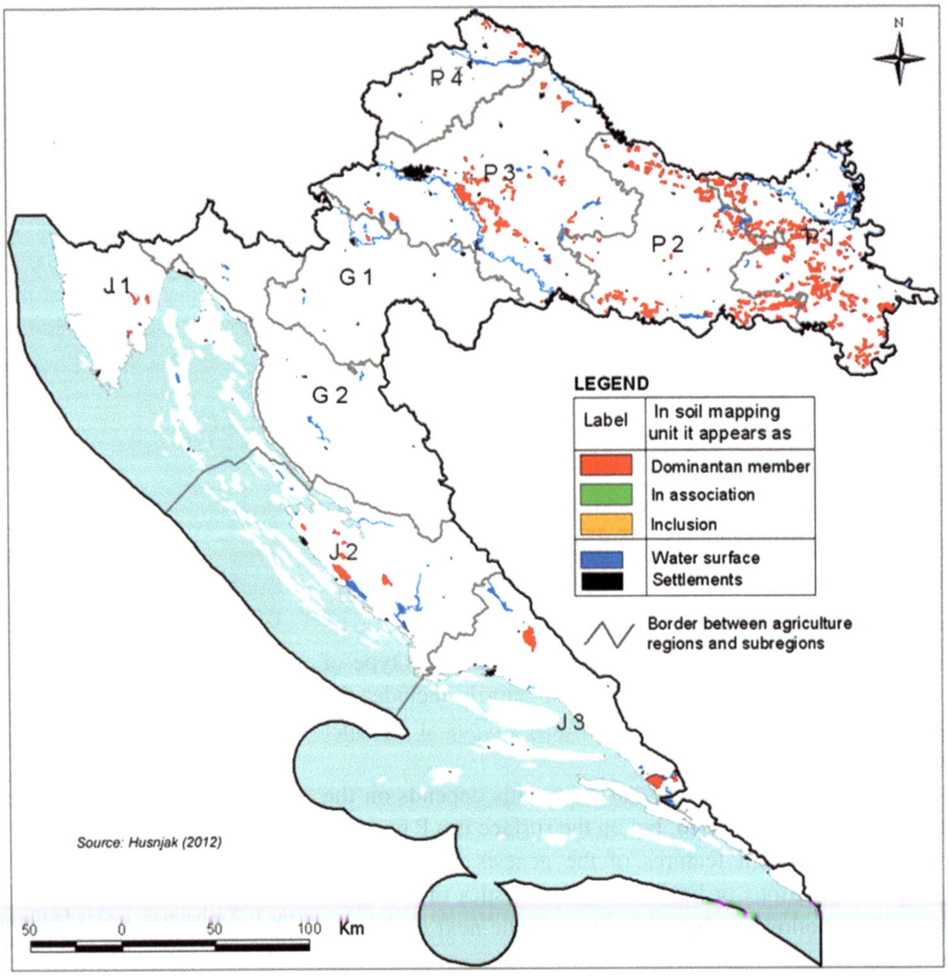

Fig. 6.27 Distribution of drained soil in the pedosphere of Croatia

Photo 6.73 The first stage of amelioration is the installation of drainage pipes (*left*), followed by the second—a deep loosening in a direction perpendicular to the pipeline (*right*), followed by a third stage of integrated reclamation—agrotechnical reclamation: liming, humization (enrichment by humus), and ameliorative fertilization

installation, distance, with/without hydraulic filters, etc.) which obligates the contractor to realize the procedure.

Most of these projects were created by the Department of Soil Amelioration of the Faculty of Agriculture of the University of Zagreb, led by Tomić (1985, 2003); Tomić et al. (1981, 1982, 1993, 1994, 2002) (Photo 6.73).

It should be noted that in some soils, such as lighter textured soils, drainage pipes are not necessary because open channels can lower the water table to the level needed for use in intensive agriculture. An example of this is humogley in which the excavation of channels in the topographically lowest parts of the plot lowers the water level so that it does not threaten the crop (Photo 6.74).

Photo 6.74 Excavation of an open channel following the lowest topographic position of the plot in humofluvisol. The water table falls to a level which does not threaten the crop

It is unnecessary to emphasize that all prior soil features are changed in substantially drained soils, and the drained soil has no similarity with the soil from which it originated. However, endomorphological traces of anaerobic processes in the soil profile will be changed very slowly—in the far future.

Although it is a separate topic, if the system is not correctly maintained, it is possible that the soil will undergo retrogradation—returning to the prior state of hydromorphism.

Physical and chemical properties. These are very heterogeneous and of course dependent on the characteristics of the soil type in which drainage was performed. The drained lands are used for the intensive cultivation of crops. It should be noted that the highest crop yields recorded in Croatia were attained in these soils. Today, the main reason for degradation of such soil is progression of hydromorphism because of lack of maintenance of the system. The soil will continue deteriorating, while we are waiting for privatization of the land, and for it to be used in intensive cultivation of crops the procedures may need to be carried out all over again.

Guidelines for management. It is self-evident that the first form of management of drained soils is drainage system maintenance and a monitoring system for drainage pipe functions to determine and react to the silting-up of drainage pipe perforations.

6.4 Order of Halomorphic Soils

Halomorphic soils are described as salty soils, formed under the influence of salty water, rich in easily soluble salts. Within this order there are three classes, each with one type of soil. These are solonchak, solonetz, and (some authors) add solonchak–solonetz.

Halomorphic soils in Croatia are very limited, symbolically represented over an area slightly larger than 500 ha. We found them around Tovarnik (Bogunović et al. 1983, 1984) and in the woods of Spacva (Vrankovic and Bašic 1975), Valpovo, Slatina, and under the influence of seawater salinization in Istria near the mouth of the Dragonja, Mirna, Rasa, and Neretva Rivers as well as around the Vrana Lake near Biograd. These soils are more interesting as a natural phenomenon because of their specific halophyte vegetation than as productive soils.

Romić and Tomić (1997), Romić and Romić (1997), Romić et al. (2008) examined some aspects of soil salinization of soils the valley of the Neretva and Vrana basin, resulting in very valuable guidance for exploitation of soils at risk of salinization.

6.4.1 Class I: Salty Soils

6.4.1.1 Solontshak

The name is of Russian origin, and signifies a salty soil. It is soil with more than 1 % easily soluble salts, mainly chlorides and sulfates, or with more than 0.7 % of soda (sodium carbonate) in the soil profile. In the dry part of the year following the flushing of salt on the surface of the soil profile or on the structural aggregate surface appear white efflorescent crusts of easily soluble salts. It has a profile of A_{sa} G or A_{sa} C Gr without textural differentiation.

Salinization can be a primary or secondary process, and it can occur in hydromorphic soils—fluvisol, semigley, and humogley. Primary salinization is created by salty water ascending to the soil surface or from the spray of seawater during periods of high seas in the coastal zone. Secondary salinization is caused by long-term irrigation using water-containing salt, or if irrigation raises the level of groundwater rich in easily soluble salts. Similar effects are obtained by raising the water table through dams for hydroelectric power or water accumulation.

Table 6.35 Taxonomy of solontshak

Name of taxon and base of classification	
Variety (based on salt content)	Form (based on dominant salts)
For all subtypes • Moderate salty 0.52–1.0 % • Very salty 1.0–2.0 % • Extremely salty > 2.0 %	For all of varieties • Soda—$HCO_3 > 50$ % of total content of soil • Soda—sulfate $SO_4 > 50$ % of total content of soil • Chlorides—$Cl > 50$ % of total content of soil • Mixed (HCO_3, SO_4 and $Cl < 50$ %)

The soil has a symbolic distribution of 121 ha only, or 0.002 % of the land area.

Taxonomy of units below subtype is shown in Table 6.35.

Salinization by water rich in sodium ions leads to alkalization. Slightly alkaline soil is considered to contain 7–15 % Na^+ ions in the CE complex, medium alkaline has 15–40 %, and strongly alkaline more than 40 % Na^+ ions in the CE complex.

According to the type of salt it is divided into: soda, sulfate, chloride and mixed; the last-mentioned contains other salts.

Solonchak soil is of minor importance, because it has very unfavorable physical, chemical, and biological features. In the spring, when the concentration of salt in it is normally lower, we have growth of natural grass vegetation. In summer, evaporation of water increases the concentration of salt, which leads to drying vegetation. Only a few halophyte plant species can endure a high concentration of salt, and of the commercially valuable plants, chamomile (*Matricaria camomilla*) can grow under a high salt content. Productive features are dependent on the amount and type of salt; the worst are soda solonchaks. Its use for intensive production in agriculture is possible only after reclamation of solonchak.

6.4.2 Class II: Alkalized Soil

6.4.2.1 Solonetz

The name solonetz, accepted in world literature, is of Russian origin and indicates a soil with low salt content.

Table 6.36 Taxonomy of solonetz

Name of taxon and base of classification	
Variety (based on alkalization)	Form (based on dominant salts)
For all subtypes	For all varieties
Moderately alkalinized (15–40 % Na^+)	Soda (HCO_3 > 50 % of total content of salt)
Very alkalinized (40–70 % Na^+)	Sulfates (SO_4 > 50 % of total content of salt)
Extremely alkalinized (>70 % Na^+)	Chlorides (Cl > 50 % of total content of salt)
	Mixed (HCO_3, SO_4, Cl < 50 % of total content of salt)

It is soil with a molic or ochric A horizon which is the eluvial horizon, from which the clay moves to an argiluvic B_t horizon with a specific and columnar structure. The horizon contains no salts, as they are moved by leaching to the B_t horizon or deeper. It occurs under conditions of a somewhat deeper saline groundwater (150–300 cm), the water table of which varies, so that there is alternating salinization in the summer and desalination—flushing of salts—in the winter period, when the salt concentration in the water is low.

Its distribution is limited—411 ha only, or 0.007 % of the land area. Table 6.36 shows the taxonomy of solonetz.

According to some authors, solonetz occurs by desalination of solonchak. The presence of Na^+ ions causes the peptization of colloids (clay and humus), so they are able in peptizated form to translocate downwards. They have a low production value and they require ameliorative interventions.

Reclamation of halomorphic soils primarily seeks to protect the soil from salty water and chemical reclamation seeks removal of harmful ions such as Na^+ ions and replacement with beneficial ones like Ca^{++} ions. This is achieved by adding gypsum, which contains calcium and an SO_4^- group, which significantly reduces the otherwise high pH.

Agrotechnical reclamation methods carried out to repair the poor physical properties of halomorphic soils include deep tillage and mixing of the mass of the soil, sowing a perennial clover/grass mixture, humification, etc. Irrigation of halomorphic soils must be carried out cautiously with small doses of water, and, of course, the water for irrigation should not contain easily soluble salts.

6.5 Order of Subaquatic Soils

Subaquatic soils are formed in underwater conditions under standing water. The dominant process in the formation of these soils is sedimentation.

In Croatia, these soils are a rarity and their distribution has not been specifically studied, except in sporadic cases when needed to determine whether the deposits of silt from the bottom of lakes and wetland could be used in agriculture and forestry. In this case the focus is on the content of heavy metals, PAHs, and residues of pesticides or radionuclides. One should always be very cautious about the content of these substances in soils and more specifically subaquatic soils.

In this order there is one class with three types of soil: dy, gytja, and sapropel, the distribution of which is only 209 ha. There is no data on the properties of these soils. It can be foreseen, however, that if the tendency towards desertification continues and even increases, fishponds and other multipurpose water accumulations may dry up, meaning the bottom of these bodies would be on the surface.

The summer of 2012 was an extremely dry year. Some of the accumulations lost their water and on the surface we can see subaquatic soil, as in Photo 6.75.

Of course, it is to be expected that not only soil but also some pollutants are sedimented in this soils.

****

What to say at the end of "The story of Croatian soils"? In the title we put an inspirational motto from the presentation of Arnuldsen on the Steering Committee of ESBN—European Soil Bureau Network, a body in the role of, as it were, the "conscience of the EU" in soil management and protection.

At first, as we stressed, Croatian soil science started in the Austro-Hungarian Empire, like some kind of "predecessor of the European Union", and developed under the Yugoslavian monarchy and the former Yugoslavia. It was never backward

Photo 6.75 A long period of drought resulted in loss of water from the accumulation of Zeleno jezero—Ričica. We can see the subaquatic soil on surface, with remains of vineyards (*left*) and trees (*right*), now covered by water (photo: Jutarnji list)

Table 6.37 Classification of soil degradation in Croatia

Level of soil degradation (state)	Causes of soil degradation (driving forces)	Processes of degradation (pressures)	Effects of soil degradation (impact)
1. Weak—simply restorable (reversible)	Improper land management	Agricultural practices: soil tillage, fertilization, etc., cause degradation of physical, chemical, and biological properties of soil as components of soil fertility	Decline of soil fertility due to: compaction, destruction of soil aggregates, soil crusting, decline of humus content, acidification, decline of soil biogenity and diversity of soil micro flora, soil infection Enhancement of costs of soil tillage and land management
2. Moderate—restorable with difficulty (conditionally reversible)	Emission of pollutants	Contamination by heavy metals and toxic elements, PAH, residues of pesticides, petrochemicals, radionuclides Dry air-deposits and acid rains	Plant products non-usable for food or fodder due to mutagenic, carcinogenic and teratogenic effects on consumers Depression of plant growth, phytotoxic effects Jeopardize other ecosystems due to emission of pollutants
3. Strong—non-restorable (irreversible)	Soil translocation	Erosion by water and wind Exploitation of stone, gravel, sand, peat, brickyards Soil removal by root and tuber Soil borrowing, covering by waste, or other soil Damage by forest fire	Loss of part or whole, of soil profile, change of natural soil stratigraphy Decline of productive land area Increase of heterogeneity of soil cover Increasing of costs of farming Loss of productive land Damage of water and other ecosystems
4. Durable—soil sealing	Permanent change of primary use	Building of infrastructure and housing Water accumulations	Permanent loss of soil function in biomass production

Bašić (1994), adapted to the DPSIR concept 2009

as regards modern development. In 2013 we are expecting a formal "return" to the European family of free peoples and of course our contribution to "the joint table of knowledge" in all fields including of the field of soil science.

Compared with the previous decade, there have been remarkable changes in soil science. For a long time soil fertility was the focus of soil sciences, as the key soil property for its function in the production of organic matter in agro- and silviecosystems. The radical change was the promotion of Multifunctional Character of Agriculture and Land—MFCAL, which focuses on "non-food" functions of soil. Damage to soil generated by non-professional soil management brings environment-regulatory functions to the fore (soil as media of retention, accumulation, transformation of toxic substances, filtering, and water protection, soil as source of emission of "greenhouse gases") (Kisić 2012, Mesić 2002, Mesić et al 2005, 2006; Šestak et al. 2012; Šimunić et al. 2002, 2005; Kisić and Bašić 2011).

As a conditionally renewable natural resource, the pedosphere is a key medium of a harmonious flow of matter and energy and maintenance of harmony and equilibrium in terrestrial and semiterrestrial ecosystems, including agricultural (agro-) as well as forest (silvi-) ecosystems. Our awareness that we are living in a world where everything is interconnected is the basis of concern about climate change, its impact on soil functions and properties, and the interrelation between them as well as reflection on soil degradation and land management. Agriculture and forestry, as renewable resource-based branches of the economy with the related processing industry, is the basis of the promise of sustainable development. They use more than 90 % of Croatian and EU-27 territory. Therefore, all registered processes of soil degradation are serious reasons for collecting data on soil degradation processes. There is a database on the state of soil and processes of soil degradation, within which one has 20 years of monitoring of water erosion of pseudogley in Daruvar (Kisić et al. 1999, 2003, 2005, 2006; Kisić and Bašić 2012), 21 years of monitoring of soils around the refinery of earth gas - hydrocarbons in the central part of Drava River valley, leaching nitrates from drained agricultural soil in intensive farming (Mesić et al. 2003, 2005) soil pollution caused by refineries of earth oil (Kisić 2012; Žužul et al. 2011) mud pits of oil mining (Kisić et al. 2007, 2008), etc.

Speaking about importance of soil degradation we think that the DPSIR concept must be used to enrich the level of collected knowledge using measurable indicators of soil degradation. We in Croatia are using a classification of soil degradation, system adapted to DPSIR, which opens the door to this process (Bašić 1994, 2009; Vidaček et al. 2003; Mesić et al. 2008). This system you can see in the last table of this book—Table 6.37.

A logical consequence of the implementation of the MFCAL concept is to open the door to radical changes of soil (land) management towards restoration of degraded soils and enhancement of all soil functions, especially the productive and regulatory functions. It means revision of the farming systems currently practiced to an adaptable farming system, adapted on all levels from agricultural region to the plot. At the same time, as a future new member state Croatia has to establish its own set of minimum standards to keep the land in Good Agricultural and Environmental Condition (GAES)

In anticipation of forthcoming access to EU, Croatian agriculture continues development in the framework of the Common Agricultural Policy (CAP) based on market economy and sustainable development of rural areas.

Experience is gained by activity in ESBN (http://eusoils.jrc.it/), a body that consists of serious and respectable scientists and has published a very extended set of data on soils of Europe, Africa, Eurasia, etc. We know that there is no alternative to sustainable land/soil management and soil protection, because of its direct influence on the quantity, food security, and quality of food and food safety, on clean and healthy drinking water and on the preservation of a pleasant landscape shaped by agriculture. All of these goods are elements of favorable living conditions, and are the "driving force" of sustainable rural development. Croatian soil scientists and land users aim to see Croatia in the circle of respectable and desirable economic partners and a member of the European society of peoples. Following "the call of time," we intend to create the basic principles of economically, ecologically, and socially sustainable and acceptable soil/land management and treatment of soil as a valuable natural resource of Croatia.

References

Bašić F, Adam M, Stepančič D (1979) Pedološka karta sekcije Rovinj-2, posebno izdanje Projektnog savjeta za izr. Opće pedološke karte Hrvatske, tisak VGI, Beograd

Bašić F, Šimunić I (1985) Pedološka karta i tumač za listove Rab-1, Rab-3, Pula-2, Projektni savjet za izradu Opće pedološke karte Hrvatske, str. 43, Zagreb

Bašić F, Kisić I, Mesić M, Knežević V (1999) Smjernice gospodarenja na tlima izloženim eroziji na područjuGrada Iloka, Zbornik znanstvenog skupa povodom 100. oblj. Poljoprivredne škole - Ilok, str. 130–153, Ilok

Bašić F, Bogunović M, Božić M, Husnjak S, Jurić I, Kisić I, Mesić M, Mirošević N, Romić D, Žugec I (2001) Regionalizacija hrvatske poljoprivrede, rukopis, Agronomski fakultet Sveučilišta u Zagrebu, 274 str., Zagreb

Bašić F, Kisić I, Mesić M (2005) Potencijalna erozija sa smjernicama održivog gospodarenja i zaštite poljoprivrednih tala područja grada Zagreba, (studija), Stručna dokumentacija Zavoda za OPB Agronomskog fakulteta Sveučilišta u Zagrebu, str. 159, Zagreb

Bašić F (2007) Soils of croatia—status—problems and anticipated solutions. In: Proceedings of mediterranean conference status of mediterranean soil resources: actions needed to promote their sustainable use, Tunis, p 87–111

Bašić F, Bašić I (2007) Navodnjavanjem tala suzbiti opustinjavanje Iloka!, Gazophylacium, God. XII br. 1–2 str. 73–93, Zagreb

Bašić F, Bogunović M, Božić M, Husnjak S, Jurić I, Kisić I, Mesić M, Mirošević M, Romić D, Žugec I (2007) Regionalisation of Croatian Agriculture, Agriculturae Conspectus Scientificus 72/1: p 35–38

Bašić F (2009) Oštećenja i tehnologije zaštite tala Hrvatske–otvorena pitanja, Zbornik radova znanstvenog skupa Tehnologije zaštite tla i odlaganja otpada, Akademija tehničkih znanosti Hrvatske str. 179–203, Zadar

Bašić F, Butorac A, Mesić M, Sabolić M (1993) Aktualna pitanja erozije i smjernice konzervacije oraničnih tala Hrvatske, Poljoprivredne aktualnosti 3–4/227–249, Zagreb

Bašić F, Butorac A, Vidaček Ž, Racz Z, Ostojić Z, Bertić B (1994) Program zaštite tala Hrvatske-Inventarizacija stanja-Trajno motrenje-Informa-cijski sustav, Zavod za OPB Agro. fakulteta, str. 122, Zagreb

Bašić F (1994) Klasifikacija oštećenja tala Hrvatske, Agronomski glasnik, 3–4, 291–310, Zagreb

Bašić F, Kisić I, Mesić M (2003) Alternativna poljoprivreda na vodozaštitnim područjima Zagrebačke županije, studija. Zavod za OPB AFZ, Zagreb

Birkas M, Jolankai M, Kisić I, Stipešević B (2008) Soil tillage needs a radical change for sustainability. Agriculturae Conspectus Scientificus 73/3: 131–136

Bogunović M, Bašić F, Adam M (1983) Pedološka karta sekcije Osijek-2, posebno izdanje Projektnog savjeta za izradu OPK Hrvatske, tisak-VGI, Beograd

Bogunović M, Pavlić V, Bašić F, Šalinović I, Bašić I (1984) Pedološka karta dijela sekcije Vukovar-1,2 i 4, i Bačka Palanka-3, Pro-jektni savjet za izradu OPK Hrvatske, tisak Vojno geografskog instituta - VGI, Beograd

Bogunović M, Vidaček Ž, Racz Z, Husnjak S, Sraka M (1996) Namjenska pedološka karta RH, mjerilo 1:300 000, Agronomski fakultet Sveučilišta u Zagrebu, Zavod za pedologiju, str. 46, Zagreb

Bogunović M, Husnjak S (1999) Tla na području grada Iloka, Zbornik znanstvenog skupa povodom 100. oblj. Poljoprivredne škole – Ilok, str. 67-80, Ilok

Bogunović M, Vidaček Ž, Husnjak S (2001) Klasifikacija tla za potrebe prostornog planiranja u Hrvatskoj, Agronomski glasnik, 63, 4–5; 171–180

Butorac A (1983) Agrotehnička studija Bušetinskog Bereka—izbor mjera i način korištenja, PIK Virovotica, Zavod za OPB Agronomskog fakulteta, str. 61–84, Zagreb

Butorac A, Bašić F, Redžepović S, Vasilj Đ (1988) Investigation into the possibilities of growing Alfa-alfa *(Medicago sativa L.)* on acric luvisol of thinly covered karst, Poljoprivredna zn. smotra, vol. 33, 5–21, Zagreb

Butorac A (1993) Conservation Tillage in Eastern Europe. Lewis publishers, Conservation in temperate Agroecosystem, pp 357–374

Butorac A (1999) Opća agronomija, Sveučilišni udžbenik. Školska knjiga, str. 648, Zagreb

Durn G (2003) Terra Rossa in the mediterranean region: parent materials, composition and origin, Geologia Croatica. J Croatian Geol Surv Croatian Geol Soc 56/1(1):83–100. Zagreb

Galović L, Frechen M, Halamić J, Durn G, Romić M (2008) Loess Chronostratigraphy in Eastern Croatia: A Luminescence Dating Approach. Quaternary International 198(1/2):85–97

Gašpar I, Leaković S, Avirović G, Horak G, Lisac H, Bašić F (2009) Preliminary investigations results of plant species and substrate selection for stabilization and vegetative cover of phosphogypsum stack. In: Proceedings of scientific meeting technologies of waste disposal and soil protection, Zagreb, p 249–257

Gračanin M (1951) Pedologija – III dio – Sistematika tala, udžbenik Sveučilišta u Zagrebu, Školska knjiga, str. 298, Zagreb

Huić Babić K, Schauss K, Hai B, Sikora S, Redžepović S, Radl V, Schloter M (2008) Influence of different Sinorhizobium meliloti inocula on abundance of genes involved in nitrogen transformations in the rhizosphere of alfalfa (Medicago sativa L.), Environmental Microbiology, 10, 11, p 2922–2930

Husnjak S, Vidaček Ž, Bogunović M, Sraka M, Bensa A, Vrhovec D (2005) Zemljišni resursi Hrvatske i pogodnost tla za navodnjavanje, Agronomski fakultet, str. 37, Zagreb

Husnjak S (2012) Sistematika tala Hrvatske, Zavod za pedologiju, Sveučilište u Zagrebu, Agronomski fakultet, Zagreb

Janeković Gj (1963) Uber die Zonalitat der Boden Jugoslawiens, Pedologie XIII p 181–194

Jug D, Birkás M, Seremesic S, Stipešević B, Jug I, Žugec I, Djalovic I (2010) Status and perspectives of soil tillage in South-East Europe. In: Proceedings of International Scientific Conference CROSTRO – Croatian Soil Tillage Research Organization, Osijek, p 50–64

Jurić I, Đikić M, Petričević P, Balen J (2000) Ishodište Pitagorinog poučka u Vučedolskoj kulturi na prostoru Hrvatske, Društvena istraživanja, vol. 9 (2–3), 347–365, Zagreb

Jurić I (2002) Počeci poljoprivredne proizvodnje na hrvatskom povijesnom području, Agriculture Conspectus Scientificus, vol. 67 (3–4), str. 181–193

Jurić I (2011) Podrijetlo Hrvata – genetički dokazi autohtonosti, Nova stvarnost, Institut Ivo Pilar, str. 181–194, Zagreb

Kisić I, Bašić F, Butorac A, Nestroy O, Marušić J, Mesić M, Sabolić M, Petraš J (1999), Zaštita tla od erozije s motrišta održivog gospodarenja tlom, Hrvatske vode, br. 26., str. 15–26

Kisić I, Bašić F, Mesić M (2002) Soil damages in Croatia and the actual problems in its protection. Assessment of the quality of contaminated soils and sites in central and Eastern European Countries (CEEC) and new independent states (NIS), str. 21–25. Sofia

Kisić I, Bašić F, Nestroy O, Mesić M (2003) Investigation of soil erosion under different tillage systems in central Croatia, Soil in the Environment, 26

Kisić I, Bašić F, Butorac A, Mesić M, Nestroy O, Sabolić M (2005). Erozija tla vodom pri različitim načinima obrade, sveuč. priručnik, knjiga, posebno izdanje Agronomskog fak. Sveuč. u Zagrebu, str. 95, Zagreb

Kisić I, Bašić F, Mesić M, Butorac A, Nestroy O (2006) Water Erosion under different crops in Croatia, 17th ISTRO (International Soil Tillage Research Organisation) Conference. Kiel, Germany

Kisić I, Bašić F, Brkić V, Mesić M, Mesić S (2007) Soil contamination at surroundings of oil-gas wells in the region County of Koprivnica-Krizevci. Cereal Res Communic 35(2):613–617

Kisić I, Mesić S, Bašić F, Brkić V, Mesić H, Sajko K, Zgorelec Ž, Jurišić A (2008) Growing of crops on soils contaminated by petrol hydrocarbons, Eurosoil, Vienna, p 77–78

Kisić I, Bašić F (2012) Influence of tillage on soil erosion, Impact of tillage and fertilization on probable climate threats in Hungary and Croatia. In: Birkas M, Mesić M (ed) Special issue of sent Istvan University Godollo and University of Zagreb, Faculty of Agriculture, Hungarian and Croatian intergovernmental S&T cooperation 2010–2011, Godollo, p 89–101

Kisić I (2012), Oštećenje i sanacija tla, udžbenik, Sveučilište u Zagrebu, str. 234, Zagreb

Kisić I, Bašić F, (2011) Imppact of tillage and fertilization on probable climate threats in Hungary and Croatia, soil vulnerability and protection. In: Birkas M; Mesić M (eds) Szent Istvan University Press, Godollo, Str. 89–100

Kovačević P, Kalinić M, Pavlić V, Bogunović M (1972) Tla Gornje Posavine, Institut za pedologiju i tehnologiju tla Poljoprivrednog fakulteta Sveučilišta u Zagrebu

Leaković S, Gašpar I, Bašić F, Horak R, Vukadin R, Avirović G, Stojić B, LisacH, Vidalin I, Matković I (2012) Postupno ozelenjivanje odlagališta fosfogipsa, Završno izvješće o rezultatima istraživanja (2009–2012), s prijedlogom izbora biljnih vrsta i supstrata za izradu izvedbenog projekta postupnog ozelenjivanja odlagališta fosfogipsa, Petrokemija d.d., str. 47, Kutina

Matić S, Tomić F, Anić I (2012) Šume, tla i vode - neprocjenjiva prirodna bogatstva Hrvatske, Zbornik savjetovanja Hrvatska akademija znanosti i umjetnosti (ur Anić I.), str. 125, Zagreb

Miloš B (1983) Tla sekcije Knin 3, Projektni savjet za izradu OPK Hrvatske 1:50 000, Split

Miloš B (1985) Tla sekcije Vis 1, 2, 3 i 4 Projektni savjet za izradu OPK Hrvatske 1:50 000, Split

Martinović J, Vranković A, Pernar N (1990) Neke pedološke karakteristike fitoklimatskih područja Velebita. Glasnik za šumske pokuse 26, s. 319–329, Zagreb

Martinović J, (2000) Tla u Hrvatskoj, Monografija, Državna uprava za zaštitu prirode i okoliša, str. 269, Zagreb

Pernar N (1997) Some relationships between cambisol over and vegetation in the west Croatian karst region. In: International scientific conference "Forest Wood Environment 97", 8. 11.IX. 1997. U Zborniku Sekcije 1 "Forest Ecology and itegrated Forest Protection, s. 11–19. Zvolen

Mayer B (1984) Pedološka istraživanja površina predviđenih za osnivanje šumskih kultura. Radovi, Izvanredno izd. 2, str. 57–61

Mayer B, Bašić F, Bogunović M (1981) Pedološka karta sekcije Karlovac-3, posebno izdanje Projektnog savjeta za izradu OPK Hrvatske, tisak VGI, Beograd

Mayer B (1992) Šumska tla Republike Hrvatske pri kraju XX stoljeća, monografija Šume u Hrvatskoj. Šum. fakultet Zagreb - JP "HŠ", str. 19–32. Zagreb

Mesić H, Bakšić D, Bašić F, Čidić A, Durn G, Husnjak S, Kisić I, Klaić D, Komesarović B, Mesić M, Miko S, Mileusnić M, Nakić Z, Pernar N, Pilaš I, Romić D, Vrbek B, Zgorelec Ž (2008a) (manual book) Croatian soil monitoring programme. Croatian Environment Agency, Zagreb

Mesić M (2002) Procjena stanja, uzroka i veličine pritiska poljoprivrede na vodne resurse i more na području RH, Poljoprivredne površine i potrošnja gnojiva, Hrv.vode, str. 1–21, Zagreb

Mesić M, Bašić F, Kisić I, Butorac A, Gašpar I (2003) Utjecaj gnojidbe mineralnim dušikom na sadržaj nitrata u tlu i na koncentraciju NO_3–N u vodi iz drenskih cijevi. Zbornik radova: 3. Hrvatska konferencija o vodama – Hrvatske vode u 21. Stoljeću, str. 359–368, Zagreb

Mesić M, Bašić F, Hamel D, Husnjak S, Kisić I, Konjačić M, Mioč B, Pavić V, Pejić I, Prpić Z, Sajko K, Vađić Ž, Vnučec I, Vuković I (2005) Izvješće o stanju okoliša u Republici Hrvatskoj. Poglavlje Poljoprivreda, Fond str. dokumentacije Zavoda za OPB, str. 109, Agr. Fak. Zagreb

Mesić M, Parlov FS, Bašić F, Kisić I, Turšić I, Zgorelec Ž, Sajko K, Vuković I (2006) Agricultural emissions of greenhouse gases in Croatia. In: Proceedings of workshop on agriculture air quality, state of the science, Washington, p 910–913

Mesić H, Bakšić D, Bašić F, Čidić A, Durn G, Husnjak S, Kisić I, Klaić D, Komesarović B, Mesić M, Miko S, Mileusnić M, Nakić Z, Pernar N, Pilaš I, Romić D, Vrbek B, Zgorelec Ž (2008) Program trajnog motrenja tala Hrvatske. Projekt, Izrada programa trajnog motrenja tala Hrvatske s pilot projektom, Life Third Countr. LIFE05 TCY/CRO/000105, str. 131, Zagreb

Resulović H, Čustović H, Čengić I (2008) Sistematika tla/zemljišta, Univerzitet u Sarajevu, str. 234, Sarajevo

Romić D, Tomić F (1997) Značajke vode Vranskog jezera u Dalmaciji kao kriteriji procjene pogodnosti za natapanje, Priručnik za hidrotehničke melioracije, II Kolo, Navodnjavanje, Knjiga 6, Kvaliteta i raspoloživost vode za natapanje, Građevinski fakultet Sveučilišta u Rijeci, str. 245–258, Rijeka
Romić D, Ondrašek G, Romić M, Borošić J, Vranješ M, Petošić D (2008) Salinity and irrigation method affect crop yield and soil quality in watermelon *(Citrullus lanatus L.)* growing. Irrig Drainage 57(4):463–469
Romić D, Romić M (1997) Ratio of salt content added by irrigation with saline water and percolated from the root zone. In: Proceedings of the international conference on water management, salinity and pollution control towards sustainable irrigation in the Mediterranean region, vol. 4, Bari, p 275–284
Stritar A (1966) Mjesto vriština u klasifikaciji zemljišta Jugoslavije, Zemljište i biljka XV 1, str. 43–52, Beograd
Stebut A (1949, 1950, 1953) Agropedologija I, II i III dio, Beograd
Šalinović I, Bašić F, Bašić I, Matun M (1999) Pokazatelji erozije tala grada Iloka, Zbornik znanstvenog skupa u povodu 100. obljetnice Poljoprivredne škole—Ilok, str. 91–118, Ilok
Šestak I, Bilandžija D, Mesić M, Gajić-Čapka M (2012) Evaluation of conditions with possible influence on agricultural practices in Croatia. In: Birkas M, Mesić M (eds.) Impact of tillage and fertilization on probable climate threats in Hungary and Croatia, Sent Istvan University Godollo and University of Zagreb, FAgric., Hung. and Croatian intergovernmental S&T coop. 2010–2011, Godollo, p 25–37
Šimunić I, Tomić F, Ostojić Z (2002) Concentration and leaching of atrazine into drainage water in Gleyic Podzoluvisol. Rostlinna Vy'roba'. 48(4):167–174. Prague
Šimunić I, Tomić F, Čoga L, Šoštarić J (2005) Influence of drainage systems on contamination of drainage water and on maize yield. Die Bodenkultur, Austrian J Agric Res (*u tisku*), Vienna
Škorić A i sur (1977) Tla Slavonije i Baranje, monografija, Projektni savjet za izradu Pedološke karte Republike Hrvatske, str. 256, Zagreb
Škorić A i sur (1987) Pedosfera Istre, monografija, Projektni savjet za izradu pedološke karte Republike Hrvatske,str. 92, Zagreb
Škorić A i sur (2003) Tla gorske Hrvatske, monografija, Ministarstvo zaštite okoliša i prostornog uređenjag fakulteta, rukopis, str. 169, Zagreb
Špoljar A, Kušec V, Kamenjak D, Kvaternjak I, Peremin-Volf T (2006) Promjene pedoloških značajki đurđevačkih pijesaka uzrokovane revitalizacijom, Agronomski glasnik br. 3, str. 181–198
Tomić F, Marinčić I, Šalinović I, Čurlin V, Ramljak T, Pejaković B (1981) Vodni režim i mogućnost daljnjih melioracija krških polja u Dalmaciji, Agronomski glasnik, br. 5–6, str. 583–606, Zagreb
Tomić F, Levaković F, Medić M (1982) Primjena cijene drenaže na površinama PIK Vinkovci, Zemljište i biljka, vol. 32 (1), str. 85–94, Beograd
Tomić F (1985) Primjena detaljne odvodnje u Hrvatskoj, Vodoprivreda, No 2/3, str. 107–113, Beograd
Tomić F, Marušić J, Buntić Z (1993) Uređenje poljoprivrednih površina u Hrvatskoj, Hrvatske vode, br. 1, str. 51–60, Zagreb
Tomić F, Šimunić I, Petošić (1994) Djelatnost različitih sustava detaljne odvodnje cijevnom drenažom, na teškom pseudoglej-glejnom tlu srednje Posavine, Agronomski glasnik, vol. 56 (1/2), str. 159–175, Zagreb
Tomić F, Šimunić I, Petošić D, Romić D, Šatović Z (2002) Effect of drainpipe spacing on the yield of field crops grown on hydromeliorated soil, Poljoprivredna znanstvena smotra, 67 (2), str. 101–105, Zagreb
Tomić F (2003) Razvojne mogućnosti melioracijskih sustava u Hrvatskoj, Hrvatske vode, br. 45, str. 375–380, Zagreb
Vidaček Ž, Racz Z, Bašić F (1993) The state, current Activities and future plans for soil monitoring systems in Croatia, FAO/ECE working party on relations between agriculture and environment. In: International workshop on harmonization of soil monitoring systems, Budapest, p 65–77
Vidaček Ž (1998) Gospodarenje melioracijskim sustavima odvodnje I natapanja, udžbenik Sveučilišta u Zagrebu, str. 217, Zagreb
Vidaček Ž, Racz Z, Crncan I, Milovic B (2003) Ucinkovitost hidro i agromelioracija u slivu Vucice iKarašice. Stanje i održivi razvoj hidromelioracijskih sustava u Hrvatskoj, Hrvatsko društvo za odvodnju inavodnjavanje, Zagreb
Vrankovic A, Bašic F (1975) Tumac pedološke karte za list Senj-2, spisi Projektnog savjeta OPK Hrvatske,str. 45, Zagreb
Vrbek B (2003) Svojstva tala šume hrasta lužnjaka i običnoga graba *(Carpino betuli Quercetum roboris Ht. 1938)* Pokupskog bazena, Česme i Repaša, Radovi, 38 2, str. 177–194
Vrbek B, Pilaš I, Pernar N, Dubravac T, Bakšić D, Medak J, Jakovljević T (2009) Heavy metals in lysimetric solution of pseudogley soils in the Kupa and Česma river areas. Periodicum Biologorum 111(4):419–426
Vrbek B, Pilaš I, Pernar N (2011) Observed climate change in Croatia and its impact on hydrology of lowlands. In: Forest management and the water cycle, Heidelberg, London, p 141–162
Vukelić J, Rauš Đ (1998) Šumarska fitocenologija, udžbenik Sveučilišta u Zagrebu, str. 310, Zagreb
Vukelić J, Mikac S, Baričević D, Bakšić D, Rosavec R (2008) Šumska staništa i šumske zajednice u Hrvatskoj, nacionalna ekološka mreža, monografija, Državni zavod za zaštitu prirode, st. 263, Zagreb
Žužul S, Zgorelec Ž, Bašić F, Kisić I, Mesić M, Vađić V, Orct T (2011) Arsenic in air and soil in the vicinity of the central gas station Molve, Croatia. Bull Environ Contam Toxicol 86:501–505

Conferences and Other Sources

Quo vadis agricultura (1997) HRSG; Absolentenverband der diplomingenieure fur Landwirtschaft, BOKU, s. 119, Wien
World reference base for soil resources (2006) World soil resources reports 103. In: Food and agriculture organization of the united nations, Rome, p 128
Zakonska regulativa u ekološkoj poljoprivredi Republike Hrvatske, posebno izdanje Ministarstva poljoprivrede i šumarstva RH, str. 200, Zagreb, 2002

Websites

European Soil Portal http://eusoils.jrc.ec.europa.eu/
Life cycle thinking, http://lct.jrc.ec.europa.eu/
Enviromental Protection http://www.zastita-prirode.hr/
Worldometers http://www.worldometers.info/
Laws and regulations http://www.nn.hr/

Curriculum Vitae

Dr. sc. Ferdo Bašić

WORLD FOOD SUMMIT
Rome, 13-17 November 1996

The author was born on 1 January 1945 in Broćanac, Posušje—Bosnia and Herzegovina. He attended primary and secondary school in Ilok—Croatia, where his family migrated. After 2 years of working on agricultural farms he studied at Faculty of Agriculture University of Zagreb, from where he graduated in 1970 and began his career. His scientific activity lasted for 43 years within the area of Biotechnology, in the field of Agriculture, and in the scientific branch of Ecology and Environment. For his postgraduate study on Soil Science he attended the same faculty (1970–1974), however, his doctoral thesis on Soil Science he defended at Faculty of Agriculture University of Sarajevo (1984).

He started as assistant at Faculty of Agriculture, University of Zagreb (1970–1972) in the Department of Soil Science, where he held terrain and laboratory training in courses on soil science, and continued at the Institute of Soil Science and Technology of Soil—Zagreb (1972–1973) as an assistant on soil genesis and soil mapping within the project of General Soil Map of Croatia in the scale 1:50 000 till finish of the project in 1986. Then followed 11 years of work (1973–1984) at the oldest Higher Agricultural College in Southeastern Europe—in Križevci in research and teaching of courses: Soil Science and Soil Amelioration. To improve teaching in higher education, in 1976, he published a textbook on soil science which already in 1982 was experiencing the second edition. This textbook defined the longer term frameworks of knowledge about soil that students of higher education in agriculture should acquire.

His career continued at the University of Zagreb, Faculty of Agriculture, in the Department of General Agronomy (1984–2010, when retired), where he was teaching General Agronomy at the graduate level. In the reformed studies in agronomy, he introduced the courses Ecological Agriculture

F. Bašić, *The Soils of Croatia*, World Soils Book Series,
DOI: 10.1007/978-94-007-5815-5, © Springer Science+Business Media Dordrecht 2013

and Soil and Water Protection, for which students showed interest and in the "Bologna transformation", which grew into a separate study on Agroecology and Ecological Agriculture, the first of its kind in this part of Europe. The Ecology of Faculty of Science University of Zagreb held a course in Agroecology as a graduate study.

Dr. F. Bašić, prof. em. introduced the concept of sustainable land management in the course Sustainable Land Management and Soil Protection and Remediation at some postgraduate studies in the Universities of Zagreb, Sarajevo, Mostar, and Split.

The author was very active in research. Chronologically, he was the first to work on the monumental project of Croatian soil scientists' Croatian General Soil Map at 1:50,000, which examines the origins, major features, and spatial distribution of over 600,000 ha of soils of Croatian pedosphere. This work marked the start of his scientific work on other topics, such as land management in agriculture, amelioration of the physical, chemical, and biological properties of soils, farming systems, and the study of the damage and effective protection of the soil as a great national treasure, in which he engaged in a well organized manner. He chaired the Department of General Agronomy, which originated in the tradition of "The Zagreb agroecological school of Mihovil Gračanin", which is known to consistently have well-known professors such as Mihalic and Butorac. He participated in over 60 scientific and professional domestic and international projects, as the head of numerous domestic and two international scientific projects, as well as on more than 80 international scientific meetings, workshops, congresses etc., worldwide. Being very communicative, open to cooperation, and team-working oriented, he developed fruitful collaboration with colleagues at home and abroad, which is illustrated in the data on the list of co-authors of his publications, where there are as much as 153 co-authors, of which 15 are from Europe, USA, and Africa.

The width of opus of his activities illustrates the numerical data on his publication. Prof. Basic is the author or co-author of 750 written publications belonging to different categories:

- Scientific papers in journals (73).
- Scientific papers revised and published in proceedings of international scientific meetings on all continents (89), papers published in proceedings of domestic meetings (56), summaries of international meetings (16), and summaries of domestic meetings (57).
- The author or co-author of 20 books, printed editions of General soil map 1:50 000 (12), chapters in the book (31), editor of five books, and co-author of two monographs.
- Invited presentations—at international meetings (12), at domestic meetings (14).

The author had very fruitful professional cooperation with farms, companies, state agencies and bodies, etc.

- Of special value are his 370 professional papers which are: studies in international cooperation (22), studies in domestic projects (66), permanent monitoring of soils (15), damage and reclamation of soils damaged by petroleum engineering, environmental incidents petrochemicals, and soils of abandoned oil fields; mud pits for gas and/or oil wells (167), soil refinery plant (4), a study of the environmental impact of mining oil facilities in the country (7), the possibility of injection of waste into abandoned wells (5), permanently abandoned wells and fields (2), a ruptured pipeline (10), the oil spill (4), etc.
- Expertise on agricultural soils (15), projects on soil amelioration (13).
- Annual reports of international projects (4), domestic projects (29).
- Reports on the work in Commission for institutional accreditation of higher school, study programme accreditation (7).

Although his 43-year scientific activities work is very rich and wide, his merits in particular may be considered in;

- issues of sustainability of soil management, soil conservation and sustainable management systems, and ecological agriculture,
- reaffirmation of crop sequence in arable farming,
- classification of soil damages/degradation,
- regionalization of Croatian agriculture,
- initiating and drafting the regulations on soil protection from pollution (NN March 1992), and defined maximal tolerant content of heavy metals and organic pollutants in the soil.

Active participation in more than 80 international scientific and expert meetings, conferences, round tables, or symposia with topics connected with land and water management, sustainable agriculture, and soil protection held on all the continents. In 1989, he was defined as head of a scientific project of 4-years' investigation of soil erosion on lateritic–ferralitic soils of the Ethiopian highland in Wollega province within a joint project of the former Yugoslavia and Republic of Ethiopia. Within this project more than 30,000 ha of virgin land was cleared and cultivated.

As the decision of the Croatian Government, in 1996 he was appointed the coordinator of the National Report on the state of food security for the "World Food Summit—Rome 1996", where over 80 reputed scientists had gathered, who made a national report entitled "Croatian Agriculture at the crossroads", and participated in the Summit as a member of the state delegation of Croatia in the role of editors and authors of the National Report.

The author is a member of scientific and professional domestic associations; Croatian Society of Soil Science, as well as The International Union of Soil Science, The International Soil Tillage Research Organization, and European Society of Soil Conservation.

Prof. Bašić executed different professional functions as:

- The representative of Croatia in Working group for Permanent Soil monitoring in Alps, Alps-Adria, and Danube river basin countries—BDF (BodenDauernbeobachtungs Fläche).
- National consultant for land resources evaluation of Croatia, within FAO project Agricultural Sector Review.
- From 2003 to 2010—Member of European Soil Bureau Network—ESBN, since 2009 also member of the Steering Committee, which is otherwise a consultative body of the European Commission (European Commission—EC) for soil management and soil protection guideline as a component of the Common Agricultural Policy—CAP.
- From 1997 to the present—one of the founders and currently serves on the editorial board of the reputable online Journal of Central European Agriculture (JCEA).
- Performed the role of chief editor Agriculturae conspectus scientificus—ACS Agricultural Scientific Review, the oldest and most respected scientific journals in the area of agriculture.
- Head of the Management council of the College of Agriculture—Križevci.
- As president of a branch of Croatian Chamber of Commerce "Eco-food", he initiated and recommended different activities for development of ecological agriculture, especially a legislative base for this orientation in agriculture.
- Deputy dean for scientific affairs of Faculty of Agriculture University of Zagreb, Head of Board for international affairs for two terms; 1993–1995 and 1995–1997.
- Appointed Honorary Advisor of the University of St. Stephen in Godollo—Hungary.
- Winner of the Annual State Science Award for 2003.
- Member of the Committee for election in academic titles in agriculture, forestry, and biotechnology for two terms.
- Member of the Working Group to prepare the National Drought Mitigation and prevention of soil damage to Croatia and author of the first national reports.
- Winner of the Lifetime Achievement Award for the year 2009.
- From 1 October 2010 retired and elected as honorable Professor emeritus.

All in all, as a reputed scientist oriented toward Sustainable development, he influenced the orientation of Croatian agriculture in a (ecologically, socially, and economically) sustainable way adapted to agroecological conditions in different agricultural regions of Croatia. In this way, he made a deep and recognizable mark in the agricultural sciences of Croatia.

e-mail: ferdo.basic1@gmail.com
website: www.ferdobasic.info

Index

F. Bašić, *The Soils of Croatia*, World Soils Book Series,
DOI: 10.1007/978-94-007-5815-5, © Springer Science+Business Media Dordrecht 2013

MIX
Papier aus verantwortungsvollen Quellen
Paper from responsible sources
FSC® C105338

If you have any concerns about our products, you can contact us on
ProductSafety@springernature.com

In case Publisher is established outside the EU, the EU authorized representative is:
Springer Nature Customer Service Center GmbH
Europaplatz 3, 69115 Heidelberg, Germany

Printed by Libri Plureos GmbH
in Hamburg, Germany